Hybrid Modeling in Process Industries

Hybrid Modeling in Process Industries

Edited by
Jarka Glassey
Moritz von Stosch

CRC Press
Taylor & Francis Group
Boca Raton London New York

CRC Press is an imprint of the
Taylor & Francis Group, an **informa** business

CRC Press
Taylor & Francis Group
6000 Broken Sound Parkway NW, Suite 300
Boca Raton, FL 33487-2742

First issued in paperback 2020

© 2018 by Taylor & Francis Group, LLC
CRC Press is an imprint of Taylor & Francis Group, an Informa business

No claim to original U.S. Government works

ISBN-13: 978-0-367-57222-8 (pbk)
ISBN-13: 978-1-4987-4086-9 (hbk)

Contents

Preface

Hybrid modeling evolved from the field of neural networks and was first reported in 1992 by Psichogios and Ungar, Johansen and Foss, Kramer et al., and Su et al. The central idea of Psichogios and Ungar was to a priori structure the neural network model through the use of first-principle knowledge. The result was that when trained with the same amount of available process data, the hybrid model was capable of predicting the process states better, was able to interpolate and extrapolate (mostly) more accurately, and was easier to interpret than pure neural networks.

In 1994 Thompson and Kramer introduced the terminology of parametric and nonparametric modeling structures to the hybrid modeling field. The parametric part of the hybrid model is a priori determined on the basis of knowledge about the system, whereas the nonparametric model parts are determined exclusively from data. The utilization of the term *hybrid semiparametric modeling* evolved from this classification to designate the type of hybrid models covered in this book, clearly distinguishing this approach from other modeling approaches that could be named *hybrid*, such as hybrid systems (continuous discrete), gray-box modeling, multi-scale modeling, block models, and others.

In the past 20 or so years, hybrid modeling has become a mature modeling methodology, with 5 to 20 articles per year being published in many different international peer-reviewed journals. In more than 200 articles on this topic, the key advantages have repeatedly been highlighted. They include better estimation/prediction performance, better inter- and extrapolation properties, less requirements on process data, and increased interpretability among other, more application-specific advantages reported in individual contributions. These key advantages are widely acknowledged among researchers in academia but also across process industries. Indeed, many of the modeling advantages give rise to an improved process operation performance, but the complexity of hybrid models—and limited education in this area and partial knowledge—appears to hinder the routine utilization of hybrid modeling methodology.

In an attempt to promote the adoption of this approach in the industries, the first hybrid modeling summer school was held at Lisbon, Portugal, in 2013, followed by a European expert meeting on "Hybrid Modeling for PAT in Biopharma: Status, Perspectives & Objectives," the outcomes of which were published in a position paper (von Stosch et al. 2014). The resulting initiatives indicated that a standard book on hybrid modeling that introduces the underlying theory/fundamentals and demonstrates practical applications in several process industries is currently missing. This book is our contribution to support the adoption and evolution of hybrid modeling methods in the process industries, allowing the exploitation of underused potential, enhancing current process operation, and improving design in an efficient way.

This book is based on contributions of leading experts in the field, and it covers both hybrid modeling fundamentals and their application. Chapters 2 to 4 cover issues such as how to develop a hybrid model, how to represent and identify unknown parts, how to enhance the model quality by design of experiments, and how to compare different hybrid models, together with tips and good practice for the efficient development of high-quality hybrid models. Chapters 5 to 8 cover the utilization of hybrid modeling for typical process operation and design applications in industries such as chemical, petrochemical, biochemical, food, and pharmaceutical process engineering. This book should therefore be of interest to researchers investigating the fundamental aspects of this modeling approach and those looking for more effective process modeling for monitoring, process development, and optimization within their fields.

MATLAB® is a registered trademark of The MathWorks, Inc. For product information, please contact:

The MathWorks, Inc.
3 Apple Hill Drive
Natick, MA 01760-2098 USA
Tel: 508 647 7000
Fax: 508-647-7001
E-mail: info@mathworks.com
Web: www.mathworks.com

Editors

Dr. Jarka Glassey currently works as a Professor of Chemical engineering education in the School of Engineering, Newcastle University, United Kingdom. She gained her academic qualifications in chemical engineering at the STU Bratislava, Slovakia, and PhD in biochemical process modeling at Newcastle University, United Kingdom. She is a Chartered Engineer, Fellow of the Institution of Chemical Engineers (IChemE), Rugby, UK, and currently serves on the IChemE Council. She is the Executive Vice President of the European Society of Biochemical Engineering Sciences (ESBES), and she is also the vice chair (immediate past chair) of the Modelling, Monitoring, Measurement & Control (M3C) Section of ESBES. Her research interests are particularly in the areas of bioprocess modeling, monitoring, whole process development, and optimization. She has published extensively in this area and over the years collaborated with a wide range of industrial partners in real-life bioprocess development and modeling applications. Currently she is coordinating a large European academic and industrial consortium carrying out research and training early career researchers in the use of the QbD, PAT and hybrid modeling approaches within biopharma industry in order to speed up the process development and reduce the lead times from discovery to full scale manufacture.

Dr. Moritz von Stosch joined the Technical Research and Development Department of GSK Vaccines in the beginning of 2017. Until then, he had worked as a Lecturer in Chemical Engineering at the School of Chemical Engineering and Advanced Materials, Newcastle University, Newcastle upon Tyne, UK, and he also was the team leader of HybPAT, a spin-off initiative with the aim of providing hybrid modeling solutions for an efficient implementation of PAT. In 2011 he earned his PhD at the Faculty of Engineering of the University of Porto, Porto, Portugal. He was awarded his diploma in engineering from the RWTH-Aachen University in Germany. Moritz von Stosch is a leading expert on hybrid modeling methods and their application to bioprocess problems. He is, for example, the author of more than ten publications on hybrid modeling and coauthor of several others. In the past years, he co-organized an expert meeting on "Hybrid Modeling for QbD and PAT in Biopharma," which was integrated into the ESBES M3C panel series, and he also co-organized all three bi-annually hold hybrid modeling summer schools. In addition, he gave a number of invited talks on "Hybrid modeling for QbD and PAT" across Europe and the United States.

Contributors

Stefano Curcio
Laboratory of Transport Phenomena
and Biotechnology
Department of Computer
Engineering, Modeling,
Electronics and Systems
University of Calabria
Rende, Italy

Vytautas Galvanauskas
Kaunas University of Technology
Department of Automation
Kaunas, Lithuania

Jarka Glassey
School of Engineering
Newcastle University
Newcastle upon Tyne, United
Kingdom

Andreas Lübbert
Martin Luther University of
Halle-Wittenberg
Centre for Bioprocess Engineering
Halle, Germany

Vladimir Mahalec
Department of Chemical
Engineering
McMaster University
Ontario, Canada

Thomas Mrziglod
Bayer
Leverkusen, Germany

Rui Oliveira
LAQV-REQUIMTE
Faculty of Science and Technology
University NOVA de Lisboa
Caparica, Portugal

Rui M. C. Portela
LAQV-REQUIMTE
Faculty of Science and Technology
University NOVA de Lisboa
Caparica, Portugal

Andreas Schuppert
RWTH Aachen
Aachen, Germany

Rimvydas Simutis
Kaunas University of Technology
Department of Automation
Kaunas, Lithuania

Moritz von Stosch
School of Chemical Engineering
and Advanced Materials
Newcastle University
Newcastle upon Tyne, United
Kingdom

1

Benefits and Challenges of Hybrid Modeling in the Process Industries: An Introduction

Moritz von Stosch and Jarka Glassey

CONTENTS

1.1 An Intuitive Introduction to Hybrid Modeling

The following illustrative example provides an intuitive introduction to hybrid modeling. An open water tank is filled with water from a pipe, and a valve regulates the water flow rate ($u - L/h$) through the pipe (see Figure 1.1). The level sensor is used to measure the tank's water level, h, at a specified time, t_f. The objective is to develop a model that can predict the water level at time $h(t_f)$ for different water flow rates u.

A statistician (assumed to have no knowledge about the underlying physical system) would likely adopt a design of experiment (DoE) paired with a response surface modeling approach, which is very systematic. The flow rate is the factor, and the water level at time t_f the response. To cover any potential nonlinear relationship between u and $h(t_f)$, typically three different flow rates (u_1, u_2, and u_3), referred to as levels in DoE, would be experimentally tested and the respective water levels at t_f recorded—$h_1(t_f)$, $h_2(t_f)$, and $h_3(t_f)$. Subsequently, a linear model could be fitted to the data (Figure 1.2),

$$h(t_f) = a \cdot u + b \qquad (1.1)$$

with model coefficients a and b. This model may predict with reasonable accuracy the water levels for any flow rate in between the tested ones.

1

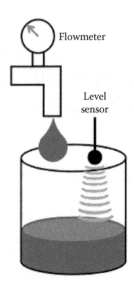

FIGURE 1.1
Water tank with flowmeter-equipped inflow and water level sensor.

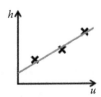

FIGURE 1.2
Height of the water tank over inflow rates. Black crosses are experimental values; gray continuous line represents the linear model.

However, is the model still valid for other final times, say $t_{f,2}$? Changes in the final time could be accounted for by introducing it as an additional factor, but this would increase the number of tests that needed to be executed. Also, until now a constant input flow rate was assumed, but what happens if u varies over time?

A process engineer would probably start the modeling exercise with the material balances. Thus, the water content at time t is equal to the content at t_0 plus the water that has been supplied to the tank in the meantime:

$$W = W(t_0) + \rho \cdot u \cdot (t - t_0) \tag{1.2}$$

with ρ the density of water, or in differential form (with $t - t_0 = \Delta t \rightarrow dt$ and $W - W(t_0) = \Delta W_{t \rightarrow t_0} \rightarrow dW$)

$$\frac{dW}{dt} = \rho \cdot u \qquad (1.3)$$

Assuming that the density is known, the water level in the tank can, in principle, be computed by taking the geometry of the tank into account, for example, for a cylindrical tank $W = r^2 \cdot \pi \cdot \rho \cdot h$. Also, changes in the water inflow during the test are explicitly accounted for. Just to check the validity of the model, the engineer would likely do one test run, where he or she would then probably find that the model prediction does not totally agree with the experimental value. Searching for an explanation of the deviation, the engineer would likely do at least another experiment and again observe a discrepancy (Figure 1.3). Since the tank is open, water will evaporate, thus resulting in a *loss* of water in the tank. Evaporation can of course be modeled, but it is not straightforward, depending on environmental conditions such as temperature, pressure, and so on. The modeling objective is not to understand evaporation, and therefore a mathematical description of evaporation could be yielded with statistical approaches.

The hybrid model in this case is hence a logical extension of the fundamental model

$$\frac{dW}{dt} = \rho \cdot u - f(T, ...) \qquad (1.4)$$

in which the unknown parts are modeled via a statistical approach, represented by $f(\cdot)$ with T, the temperature. The data required for the development of the statistical model could likely be gathered during normal process operation conditions, allowing the refinement of the hybrid model in an iterative experiment-to-experiment fashion. It would also be possible to execute a DoE with temperature and so on as factors and to use the resulting data for the development.

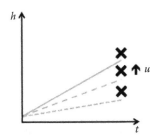

FIGURE 1.3
Height of the water tank over time. Black crosses are experimental values at different inflow rates; gray lines represent the simulation of the fundamental model at different inflow rates.

1.2 Key Properties and Challenges of Hybrid Modeling

Summarized in one sentence, the key advantage of hybrid modeling is its high benefit-to-cost ratio for modeling complex systems, predominantly a result of the incorporation of fundamental knowledge.[*] The knowledge integration gives rise to three key properties that are at the heart of all hybrid modeling benefits:

1. Hybrid models provide an improved understanding of the system, yet not all parts of the model need to be fundamentally understood, but instead can be represented by statistical modeling approaches. For instance, evaporation was modeled by a statistical model in the example, but the interplay between filling the tank and evaporation becomes clear. Also, the impact of variations in the inflow can be analyzed. The opportunity to assess the impact of dilution on the concentration profile in more complex fed-batch bioreaction or chemical reaction processes is another straightforward example. From the quality control perspective this is a very important property, as it allows an easy route-cause analysis to explain variation or failures. This property might also help when, for example, looking for process approval by agencies.

2. The requirements on the data are lower for hybrid models than for purely data-driven models,[†] both in quality and quantity. One reason is that the impact of certain factors (variables) is explicitly accounted for by the incorporated fundamental knowledge; accordingly, these factors are not incorporated into the statistical model. This is for instance the case in the water tank example, where the impact of the water inflow is directly described by the fundamental model. Another reason is that the fundamental knowledge introduces structure, which can describe, for example, the interaction between variables, reducing the need to investigate these interactions experimentally. This is a very important property as this means that a predictive model potentially could be developed from large data sets that do not inherit a lot of variation in the variables, such as big data or typical process data. Therefore hybrid modeling could potentially help overcome the curse of dimensionality, as has been studied in more detail, for instance by Fiedler and Schuppert (2008). In the water

---

[*] The term *fundamental knowledge* is used in this chapter as a representation for knowledge that stems from one or more persons' experience and is readily expressed in mathematical form, such as first-principles, mechanistic, or empirical knowledge.
[†] The term *data-driven models* is used to account for all methods that are "data driven," such as statistical modeling, chemometric modeling, and MVDA modeling, among others.

tank example, a possible interaction between temperature and water inflow may be considered, but their impact is clearly separated in the model structure; hence the interaction does not need to be assessed experimentally.

3. The extrapolation properties of hybrid models have several times been reported to be excellent (Van Can et al. 1998; Te Braake et al. 1998; Oliveira 2004), allowing reliable model predictions beyond experimentally tested process conditions—a property that is crucial for process development, optimization, and control. In the water tank example, the final time and inflow rate can be changed without loosing prediction power, extrapolating beyond tested times or flows. However, for changes in temperature that are different from the experimentally tested (observed) ones, the model will likely not predict reliably, as the data-driven model would have to predict outside the region where it was trained, a scenario for which data-driven models are known not to perform well. However, it should be noted that small changes to the model can make a big difference for its extrapolation properties. For instance, if the intention was to use the model for scaling up the water tank, one could make the evaporation surface area specific; thus the dimensions of the water tank would not affect model validity in any way. The intra- and extrapolation properties have been studied in detail by Van Can et al. (1998), and insights for scale-up focused model development have been published by Te Braake et al. (1998).

However, the integration of fundamental knowledge not only brings about the key properties, but also raises three challenges:

a. In the same way that the fundamental knowledge helps to structure the interaction and/or defines the shape of the model predictions, it determines the *prediction plane/response surface* in which the prediction can be found—the fundamental knowledge imposes an inductive bias on the model predictions (Psichogios and Ungar 1992; von Stosch et al. 2014). However, if the fundamental knowledge is incorrect, it might never be possible for the prediction to correspond to the actual system response, which in turn cannot be found on the *plane/surface*.

b. The parameters, which potentially can be found in both the fundamental and statistical models, need to be identified. Whereas standard techniques are available for purely fundamental or statistical models, only a few parameter identification methods for hybrid models have been reported (Psichogios and Ungar 1992; Kahrs and Marquardt 2008; von Stosch et al. 2011; Yang et al. 2011). The major challenge is the time required for parameter identification. It can still take a significant amount of time (up to one month) to identify the parameters that provide the best model performance, even with the computing power available these days.

c. Given the first challenge, the development of hybrid models needs to consider the *correctness* or underlying assumptions of the fundamental knowledge that should be incorporated. There are (at least) two possible strategies that one can follow: (i) integrate only the most basic knowledge, assess model performance, and then iteratively introduce more knowledge until model performance decreases; or (ii) integrate all available knowledge, analyze model performance, and try to iteratively remove what is hindering the model to perform good. However, in light of the time required for parameter identification (and also the discrimination of the best underlying data-driven model structure), both procedures might take a significant amount of time. Thus, model development will either require some experience or an analysis of a risk of fundamental knowledge incorrectness.

The key properties are obviously very appealing, and from personal experience the authors can state that the challenges, if appropriately addressed, represent no significant hindrance to the exploitation of the properties and achieving all the benefits that hybrid modeling can offer.

1.3 Benefits and Challenges of Hybrid Modeling in the Process Industries

In the process industries, modeling is not a self-fulfilling exercise, but it needs to provide a better understanding of the underlying mechanisms to improve the economics of the process or its development. Thus, the objective of the modeling exercise is to improve process understanding, enabling a rational manipulation of the process. The aim is to maximize the benefit to cost ratio. It is often difficult to exactly quantify the benefit, but an approximation of the potential might exist. However, the purpose of the model—its use for process understanding, characterization, development, monitoring, control, scale-up, or optimization—will impose certain requirements on the model to yield a good performance. The more the purpose requires a good global model description of the process and the more detailed the understanding of the underlying mechanisms must be, the greater the need for a more fundamental description (Figure 1.4). In contrast, the more local the model descriptions are or if approximations suffice, the better the preposition of data-driven models. It is evident that hybrid modeling helps bridge

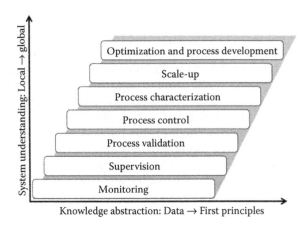

FIGURE 1.4
Classification of process design and operation disciplines in terms of system (process) understanding and level of knowledge abstraction aiming for good performance.

the gap between data-driven and fundamental modeling, at little extra cost, which is the other factor that drives the cost-benefit ratio. The modeling objective (purpose) and the knowledge sources available at the beginning of the modeling exercise determine the effort that must be dedicated to the model development, and thus the cost. In general, the less fundamental knowledge is available, the greater the requirements on process data. However, as discussed earlier for the key properties, the integration of the available fundamental knowledge with data-driven techniques can help reduce the data requirements, still yielding a model that can extrapolate sufficiently accurately. Therefore, increased benefits can be achieved at a similar cost.

Two types of challenges for the application of hybrid modeling exist: technical and human-related challenges. The technical challenges mostly stem from the limited applications of hybrid modeling in the process industries. For instance, it has not yet been reported how to validate a hybrid model reliably. Also, questions around the maintenance of hybrid models have not yet been investigated. The introduction of a commercial hybrid modeling software would go some way toward addressing these challenges. However, approaches for model maintenance and validation can likely be adopted from the fields of fundamental or data-driven modeling. Human-related challenges arise from missing expertise and experience. The development of hybrid models draws from two pools of expertise: engineering and statistical expertise. However, the curriculum of process engineers typically contains little education on statistical modeling, despite the fact that artificial intelligence is thought to have the capacity to change the process industries dramatically. Also, the education of statisticians does not comprise subjects on process/systems modeling from first principles. Coming to the job, it is hard to acquire the necessary expertise to develop hybrid models, and even more so since commercial software is currently nonexistent. Given the limited hybrid modeling experience present in the industries, assessing the risk of a hybrid modeling application is difficult. Without a clear need or a clear benefit case, there will thus be an inclination toward applying the traditional approaches even if they might come at a greater cost. At present, a community is developing in which an experience exchange on hybrid modeling topics is fostered, and this book is attempting to contribute to such an exchange. The editors and authors of this book hope that the information within will help increase the number of hybrid modeling applications in the process industries. However, it is evident that education also needs to contribute, to enable a more efficient use of the available process knowledge via hybrid modeling.

1.4 Hybrid Modeling: The Idea and Its History

Hybrid modeling is thought to have evolved from the field of neural networks around 1992 through the work of Psichogios and Ungar; Johansen and Foss; Kramer, Thompson, and Bhagat; and Su et al. The original idea was to introduce structure into the neural network models to obtain models that are more reliable, easier to interpret, and more capable of generalization and extrapolation than classical neural networks (Psichogios and Ungar 1992). In fact, these are the key properties of hybrid models summarized earlier and also repeatedly reported throughout the past 25 years (Oliveira 2004; Thompson and Kramer 1994; von Stosch et al. 2014). In 1994, hybrid models were described as a combination of nonparametric and parametric modeling structures (Thompson and Kramer 1994), which gave rise to the term *hybrid semiparametric modeling* that was promoted by von Stosch et al. (2014) to distinguish these types of models from other hybrid models. Combinations of parametric and nonparametric modeling structures can, however, be found before 1992, as for instance partially linear models (Robert et al. 2001) or index models (Ichimura 1993). The detail that probably differentiates these nonparametric/parametric models from the neural network–based hybrid models developed around 1992 is the fact that the parametric part is fixed by process knowledge in the latter approaches, whereas it emerged from statistical considerations in the earlier ones.

Around the same time as partially linear models were studied, gray-box modeling/identification emerged in the field of control theory. A first session on gray-box identification was held at the fourth IFAC Symposium on Adaptive Systems in Control and Signal Processing in 1992, and in 1994 a first special issue in *International Journal of Adaptive Control and Signal Processing* was published on the same topic (Bohlin 2006). Gray-box modeling, like hybrid modeling, is based on the idea of using prior process knowledge, together with statistical approaches, exploiting the process data (Bohlin 2006). However, the approach of gray-box modeling is understood to be broader than that of hybrid modeling in the sense that any incorporation of fundamental knowledge with data-driven techniques can be considered gray-box modeling, although only those models that combine parametric and nonparametric structures are deemed hybrid, as considered in this book. Hybrid models can be viewed as a particular class of gray-box models (von Stosch et al. 2014).

The first practical application of hybrid models for upstream bioprocess monitoring, control, and optimization was reported in 1994 (Schubert et al. 1994), and several practical applications of hybrid modeling for biochemical process operation followed. Shortly after this, hybrid modeling was also adopted by different industry branches, and hybrid approaches can now be found across various process industries (von Stosch et al. 2014). Though more than 25 years old, hybrid modeling seems to again enjoy increasing interest in recent years. For instance, in 2013 a position paper was published by the ESBES's M3C Section on the application of hybrid models in the (bio)pharmaceutical industry for the implementation of the process understanding–centered guidelines proposed in 2004 by the U.S. Food and Drug Administration (von Stosch et al. 2014). Also, at the final stages of the editing of this book, there was a call for the submission of manuscripts for the first special issue on Hybrid Modeling and Multi-Parametric Control of Bioprocesses in the *Bioengineering Journal*. With the ever-increasing amount of data that is generated, the vast amount of knowledge that is produced in all fields of research, and the increasing computer power that allows the development of larger models and the rise of the *system* research fields (e.g., process system engineering, systems biology, or systems medicine) that seek to combine knowledge into an integrated and coherent whole, it seems that hybrid modeling has a golden future.

1.5 Setting the Stage

This chapter provides an intuitive introduction to hybrid modeling, its history, and its key properties and challenges as well as the benefits and challenges of its application in the process industries. The following chapters either detail methods for the development of hybrid models or provide an overview of the hybrid modeling applications in specific process industry sectors.

In particular, Chapter 2 introduces different possible hybrid modeling structures and how these structures can be exploited when different levels of fundamental knowledge and/or amounts of process data are available. Model structure requirements for process operation and design are discussed.

Two examples illustrate possible structures—one using bootstrap aggregated dynamic hybrid models for bioreactor modeling and another exploiting a static parallel hybrid modeling structure for activity modeling of ribosome binding sites.

Chapter 3 explores the relationship between available experimental data and the validity domain of the model. The opportunity to use design of experiments (DoEs) and model-based design (optimal experimental designs) to systematically develop and extend the validity domain of the model are discussed. The approaches are contrasted and illustrated with a bioreactor example, highlighting also a model-based approach for parallel experimentation as, for example, promoted for modern high-throughput methods.

In Chapter 4, the properties of tree-structured hybrid models are addressed and the techniques for parameter identification and structure discrimination are discussed in detail. Four practical industrial examples from the chemical industries are provided: (1) optimization of a metal hydride process, (2) modeling of complex mixtures, (3) software tool design and processing properties, (4) and surrogate models for the optimization of screw extruders. The benefits of hybrid models over pure neural networks are highlighted. For example, for the optimization of the metal hydride process, the incorporation of knowledge into the hybrid model allows the reduction of the number of inputs of the data-driven model and increases the area in which reliable predictions can be made.

The application of hybrid models for monitoring and optimization of a number of different bioprocesses are presented in Chapter 5, including penicillin production, microbial surfactant production, recombinant protein production, and mammalian cell cultivations. Besides neural networks, fuzzy models constitute the nonparametric part of the hybrid models. The use of hybrid approaches with Kalman filters and extensions to the filters are presented for process monitoring. Different process control approaches are discussed, as well as their differing process understanding requirements. Finally, the potential of hybrid models for fault diagnosis and application in virtual plants (which can be implemented in all modern process control software) is highlighted.

Hybrid modeling applications in the petrochemical sector are described in Chapter 6—in particular, hybrid models for petrochemical reactors, such as a highly exothermic fixed-bed reactor, a continuous stir tank reactor, and a fixed-bed catalytic reactor as well as distillation towers. It is shown that *simple* hybrid models can be used instead of more complex rigorous mechanistic models, reducing the simulation time significantly without compromising simulation quality, which is especially interesting for on-line process optimization and optimal control.

Chapter 7 describes hybrid modeling applications for a food transformation and a food-waste valorization process—the convective drying of vegetables and enzymatic transesterification of waste olive oil glycerides for biodiesel production, respectively. A set of material, momentum, and energy balances constitute the backbone of the food drying model, whereas changes in parameters such as redness, yellowness, lightness, or microbial inactivation in the function of the operating conditions were modeled using neural networks. Describing the changes in these parameters mechanistically is particularly difficult, which makes the adoption of a hybrid modeling approach intuitive.

Chapter 8 discusses the applicability of hybrid modeling approaches in the pharmaceutical and biopharmaceutical industrial sectors. The future perspectives of the sector are briefly reviewed with reference to the regulatory initiatives driving the process development in this sector—the quality by design and process analytical technologies. Following a brief review of pharmaceutical applications of hybrid modeling in tableting and other relevant processes, the chapter then concentrates on a specific part of the sector predicted to experience the most significant growth in the near future; this includes the production of biologics, such as monoclonal antibodies (mAbs), using cell-culture cultivations. A case study is presented discussing details of model development to predict critical quality attributes and critical process parameters, including mAb yields, glycosylation profiles, and the concentrations of important metabolites throughout the cultivation.

We hope that the breadth of detailed examples, together with robust theoretical grounding of the concepts covered in this book, will enable its readers to consider or extend the use of hybrid modeling in whichever industrial or research sector they practice.

References

Bohlin, T. P. 2006. *Practical Grey-Box Process Identification: Theory and Applications, Advances in Industrial Control.* London, UK: Springer Science & Business Media.

Fiedler, B. and A. Schuppert. 2008. Local identification of scalar hybrid models with tree structure. *IMA Journal of Applied Mathematics* 73 (3): 449–476. doi: 10.1093/imamat/hxn011.

Ichimura, H. 1993. Semiparametric least squares (SLS) and weighted SLS estimation of single-index models. *Journal of Econometrics* 58 (1): 71–120. doi: 10.1016/0304-4076(93)90114-K.

Johansen, T. A. and B. A. Foss. 1992. Representing and learning unmodeled dynamics with neural network memories. *American Control Conference.* Chicago, IL: IEEE, June 24–26, 1992.

Kahrs, O. and W. Marquardt. 2008. Incremental identification of hybrid process models. *Computers and Chemical Engineering* 32 (4–5): 694–705. doi: 10.1016/j.compchemeng.2007.02.014.

Kramer, M. A., M. L. Thompson, and P. M. Bhagat. 1992. Embedding theoretical models in neural networks. *American Control Conference.* Chicago, IL: IEEE, June 24–26, 1992.

Oliveira, R. 2004. Combining first principles modelling and artificial neural networks: A general framework. *Computers and Chemical Engineering* 28 (5): 755–766. doi: 10.1016/j.compchemeng.2004.02.014.

Psichogios, D. C. and L. H. Ungar. 1992. A hybrid neural network-first principles approach to process modeling. *AIChE Journal* 38 (10): 1499–1511. doi: 10.1002/aic.690381003.

Robert, F. E., C. W. J. Granger, J. Rice, and A. Weiss. 1986. Semiparametric estimates of the relation between weather and electricity sales. In *Journal of the American Statistical Association* 81 (394): 310–320. New York: Taylor & Francis. doi: 10.1080/01621459.1986.10478274.

Schubert, J., R. Simutis, M. Dors, I. Havlik, and A. Lübbert. 1994. Bioprocess optimization and control: Application of hybrid modelling. *Journal of Biotechnology* 35 (1): 51–68. doi: 10.1016/0168-1656(94)90189-9.

Su, H. T., N. Bhat, P. A. Minderman, and T. J. McAvoy. 1992. Integrating neural networks with first principles models for dynamic modeling. *IFAC Sympos.* DYCORD, Maryland, USA.

Te Braake, H. A. B., H. J. L. van Can, and H. B. Verbruggen. 1998. Semi-mechanistic modeling of chemical processes with neural networks. *Engineering Applications of Artificial Intelligence* 11 (4): 507–515. doi: 10.1016/S0952-1976(98)00011-6.

Thompson, M. L. and M. A. Kramer. 1994. Modeling chemical processes using prior knowledge and neural networks. *AIChE Journal* 40 (8): 1328–1340. doi: 10.1002/aic.690400806.

Van Can, H. J. L., H. A. B. Te Braake, S. Dubbelman, C. Hellinga, K. C A. M. Luyben, and J. J. Heijnen. 1998. Understanding and applying the extrapolation properties of serial gray-box models. *AIChE Journal* 44 (5): 1071–1089. doi: 10.1002/aic.690440507.

von Stosch, M., S. Davy, K. Francois, V. Galvanauskas, J.-M. Hamelink, A. Luebbert, M. Mayer et al. 2014. Hybrid modeling for quality by design and PAT-benefits and challenges of applications in biopharmaceutical industry. *Biotechnology Journal* 9 (6): 719–726. doi: 10.1002/biot.201300385.

von Stosch, M., R. Oliveira, J. Peres, and S. Feyo de Azevedo. 2011. A novel identification method for hybrid (N)PLS dynamical systems with application to bioprocesses. *Expert Systems with Applications* 38 (9): 10862–10874. doi: 10.1016/j.eswa.2011.02.117.

von Stosch, M., R. Oliveira, J. Peres, and S. Feyo de Azevedo. 2014. Hybrid semi-parametric modeling in process systems engineering: Past, present and future. *Computers and Chemical Engineering* 60: 86–101. doi: 10.1016/j.compchemeng.2013.08.008.

Yang, A., E. Martin, and J. Morris. 2011. Identification of semi-parametric hybrid process models. *Computers and Chemical Engineering* 35 (1): 63–70. doi: 10.1016/j.compchemeng.2010.05.002.

2

Hybrid Model Structures for Knowledge Integration

Moritz von Stosch, Rui M. C. Portela, and Rui Oliveira

CONTENTS

2.1 Introduction

Developing a model of a physical system is essentially an exercise of translation of existing sources of knowledge into a compact mathematical representation. The availability of knowledge and the quality of said knowledge is the key determinant of the model structure and subsequent model performance. In some cases, a fundamental physical understanding exists that is best expressed in the form of a mechanistic model. For other cases, the absence of a comprehensive knowledge base is compensated for by measured process data that can best be modeled by empirical methods. For the vast majority of problems in process systems engineering or in systems biology, both types of knowledge coexist, motivating their integration into hybrid models in the sense of knowledge integration.

2.1.1 Different Types of Knowledge for Model Development

Different knowledge types can be considered for the development of models. Probably the most basic form of knowledge is qualitative and quantitative data. Statistical models can be developed from these data, capturing the data underlying relations between variables by mathematical functions. These models provide a way to understand the system better and thus help gain more *abstract* knowledge. However, the prediction/estimation quality of these models critically depends upon the quality of the data. A more abstract form of knowledge is heuristics. Heuristics typically stem from people's experience of repeating a certain process/procedure, such as when operators know at what time or under which conditions they have to change certain process parameters. These heuristics can be captured in the form of fuzzy models or expert systems, and they have been shown to be capable of describing important process features (Nascimento Lima et al. 2009; Nair and Daryapurkar 2016; Qian et al. 2003; Kramer and Eric Finch 1988). Phenomenological models describe phenomena by using mathematical functions. Though they can describe the observed behavior very well, they do not explain or capture the underlying mechanisms. This is accomplished by mechanistic models and models that are derived from first principles. These models are considered to inherit the highest level of abstraction, which means that they can describe the behavior of the system beyond the experimentally investigated conditions. First-principle knowledge—material or energy balances—is typically used as a modeling framework in engineering science. The balances can be derived from the knowledge about the system and its configuration, and they are generally valid. However, the descriptions of the transport and reaction mechanisms that are contained within these balances are frequently represented by empirical equations. Thus, these models may be referred to as hybrid models in the sense of knowledge integration, but not necessarily as hybrid semiparametric models, which are the focus of this book. This issue is further discussed next.

2.1.2 Parametric/Nonparametric Modeling Paradigms

Another approach for model classification, which stems from statistics and was first introduced into the area of hybrid modeling by Thompson and Kramer (1994), separates models into parametric and nonparametric models. The difference between parametric and nonparametric models is that the structure is inferred from data in the case of nonparametric models, whereas the structure is fixed by knowledge a priori in the case of parametric models. Both nonparametric and parametric models contain parameters. In the case of parametric models (in contrast to the nonparametric ones), the parameters have a physical/phenomenological meaning. Parametric models constitute mechanistic and phenomenological models, and they are typically derived from conservation laws (material, momentum, and energy), thermodynamic, kinetic, and/or transport laws. On the other hand, kernels, splines, wavelets (flexible structure/parameters), artificial neural networks (ANNs), and projection to latent structures (PLS) are examples of nonparametric models. The combination of parametric and nonparametric modeling approaches results in hybrid semiparametric models. This means that one part of the model has a fixed structure, while the other part has a structure that is inferred from data.

2.1.3 Why Hybrid Semiparametric Modeling?

Mechanistic modeling and data-driven modeling constitute two approaches that are different in their traits. The development of a mechanistic model can be cumbersome and/or laborious. It requires detailed knowledge about the process, but its extrapolation power is typically very good, superior to that of data-driven models. However, data-driven models are rather quickly applicable, and their development can be very systematic. They require a lot of data, and their descriptive quality is good only in the vicinity to those regions for which they were derived.

Hybrid semiparametric modeling can balance the advantages and disadvantages of strictly mechanistic and data-driven modeling. It can, in principle, lead to (1) a more rational use of available knowledge and (2) a higher benefit-to-cost ratio to solve complex problems, which is a key factor for process systems engineering.

2.2 Hybrid Semiparametric Model Structures

2.2.1 Parametric and Nonparametric Models

As a general formulation of parametric models, the expressions

$$Y = f(X, w) \qquad (2.1)$$

can be used, where Y is the output of the parametric model, X is the input to the parametric model, $f(\cdot)$ is a set of parametric functions (the structure of which is determined by a priori knowledge), and w represents a set of parameters.

A general representation of the nonparametric models is

$$Y = g(X, \omega) \qquad (2.2)$$

where:
  Y is the output of the nonparametric model
  X is the input to the nonparametric model
  $g(\cdot)$ is the set of nonparametric functions
  ω represents a set of parameters (called weights in the case of neural networks)

The structure of the nonparametric functions $g(\cdot)$ and their parameters ω are determined from data. This can be a tricky task, because (for example) neural networks can approximate any function and even describe the data of inherent noise patterns. There are two challenges that are associated with structure discrimination and parameter identification:[*] (1) which nonparametric structure should be selected and (2) when the parameter identification should be stopped.

The objective is to obtain a nonparametric model that can describe the behavior of the system for new data (data that have not been used for parameter identification or structure discrimination). With this objective in mind, the common approach to overcome the challenges is the partition of the available data into three sets: training, validation, and test sets typically containing ~70%, ~20%, and ~10% of the data, respectively. The training set is used to identify the parameters, while the validation set is used to determine when the training should be stopped—from which instance on does the nonparametric model learn the data inherent noise pattern rather than the underlying function (that is also present in the validation data). To discriminate the most suitable nonparametric model structure, systematic different possible nonparametric modeling structures are trained, and their performance on

---

[*] Parameter identification is also referred to as "training," for example, in the field of neural networks.

the training, the validation, and in particular the test data is compared. For the comparison, not only the fit of the model predictions to the experimental data is taken into account, but also the complexity of the model. According to Occam's razor, "explaining a thing no more assumptions should be made than are necessary," meaning that from two models producing a similar fit, the one with the simpler structure should be selected. Balancing model complexity and fit, criteria such as the Bayesian Information Criteria (BIC) or the Akaike's Information Criteria (AIC) can be used for selecting the model structure (Burnham and Anderson 2004). However, the search for simpler model structures can also be partially integrated into parameter identification, driving parameters that are not significantly contributing to the model fit to zero via regularization approaches (Willis and von Stosch 2016) (see also Section 2.4 or Chapter 3).

In the case where only a very small number of data points are available, the performance of the above-mentioned approaches will critically depend on the sampling of the data—which data are attributed to the training, validation, and test data. In order to overcome problems associated with small data sets, bootstrapping techniques can be used (Breiman 1996; Mevik et al. 2004) (see also example 2.3.2).

2.2.2 Semiparametric Models: Serial and Parallel Structures

Partially linear models and index models are two particular cases of parallel and serial hybrid semiparametric modeling structures, which evolved in the field of statistics. Partially linear models were first introduced in 1986 to study the impact of weather on electricity demand (Robert et al. 2001), and this model has been studied extensively since then. It combines a linear model (parametric part) with a nonlinear estimator:

$$Y_{mes} = \beta \cdot X + g(X, \omega) + \varepsilon \qquad (2.3)$$

where:
Y_{mes} is the target output
β are the coefficients of the linear model
ε are the residuals

In index models, in which identification is, for example, investigated by Ichimura (1993), the linear model is transformed by a nonparametric function

$$Y_{mes} = g(\beta \cdot X) + \varepsilon \qquad (2.4)$$

Generally, two models can be combined in three possible ways; one parallel and two serial structures are feasible. Which structure is preferred is largely a question of how much fundamental knowledge is available about the process at hand. In the case where a full fundamental (mechanistic model) is available but its predictions do not fit the process data satisfactorily, a data-driven structure can be used in parallel to improve (correct) the predictions, giving rise to a parallel hybrid modeling structure. In cases where a well-structured mechanistic model is available but parts of the process are not well understood (unknown), the data-driven model can be used to describe the unknown/poorly understood part of the process. This will result in a serial hybrid model such as the one displayed in Figure 2.1a or c. The question that distinguishes the two serial structures is whether the backbone (weighting of the different mechanistic parts) of the model is known or whether the backbone is accomplished using a data-driven model. These structures represent the major *motifs* and can be combined. Also, the serial structure displayed in Figure 2.1c and the parallel structure are very similar, as the fundamental model in the case of the parallel structure could be used as an input to a data-driven model to correct for eventual mis-predictions. In general, it is not clear which structure is preferred, and the choice of structure is therefore case dependent. In particular, this might depend on the functionality that is left within the residuals that are to be modeled by the data-driven model in the parallel case.

A major difference between the two serial structures is the requirement regarding the mechanistic knowledge. Whereas in the serial structure shown in Figure 2.1c potential mismatches (between model and plant) can be accounted for, this is only possible to a very limited degree with the structure shown in Figure 2.1b. Imagine, for instance, the following example. Two reactions, r_A and r_B, are observed and thought to be linked. If these reactions are correlated using a fixed yield—$r_A = Y_{AB} \cdot r_B$—then potential mismatches in describing the reaction r_B will be directly propagated to r_A. Also, since the yield is constant, changes in the true process yield that might be, for example, due to the process variable not accounted for directly in the model

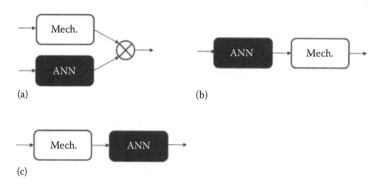

(a)　　　　　　　　　　　　　　　(b)

(c)

FIGURE 2.1
(a) Parallel hybrid model structure; (b) serial hybrid model with a nonparametric and a parametric model in series; and (c) serial hybrid model with a parametric and a nonparametric model in series. The parametric model is represented by the white box, the nonparametric model by the black box.

are not accounted for. If r_A was a function of r_B given by some data-driven model—for instance, $r_A = h(r_B, X, w)$—then the correlation is no longer fixed, and also temperature dependence could easily be integrated (temperature would be one of the variables covered in X).

Therefore, and as outlined earlier, the question of which model structure is preferred depends very much on the available knowledge and the certainty about the *correctness* of this knowledge.

2.2.3 Gating Functions

In the case where several knowledge sources are available to describe the same variable, gating functions can be used to weight the different alternatives. The simplest possible weighting is the summation or multiplication of the outputs in the parallel structure, shown in Figure 2.1, giving per se equal weight to the models. However, it is possible to use gating functions for assigning weights to the model outputs, as shown in Figure 2.2. Different approaches can be used as gating functions: for example, neural networks, experience measures, rule-based expert systems, or exponential switchers (mixture of experts). An obvious possibility is the application of an additional nonparametric model to weight the predictions of the parallel parametric and nonparametric models. Since tasks of the two nonparametric models are different, the identification of the involved parameters likely becomes easier. In the experience measure approach, the data that the model was developed on (training domain)—the experience—is used to weight the contributions of the knowledge sources. A clustering approach or radial basis function approach can be used to characterize the training domain, which then allows the assessment of the position of a new point in relation to the training domain (Simutis et al. 1997; Teixeira et al. 2006). Depending on the position of the point in relation to the training domain, the weighting of the models is established. If a new data point is close to the training domain, more weight will be given to the nonparametric model, which should perform excellently on the domain for which it was developed. If the data point moves away from the training domain, then more weight will be given to *higher-level/abstract* knowledge, which is expected to extrapolate/generalize better. This approach has been

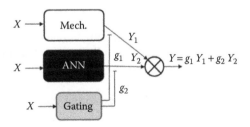

FIGURE 2.2
Parallel hybrid structure with gating function.

used for weighting different knowledge sources for the modeling of a mammalian cell cultivation (Simutis et al. 1997). In rule-based approaches, this weighting is established based on rules. The mixture of expert approach is in some sense an extension to the experience measure, but the experience domain of each model is inferred from measurements of the target (Y_{mes}) by, for example, using an expectation maximization approach. This approach has been applied successfully by Peres et al. for the modeling of a wastewater treatment plant (Peres et al. 2001, 2008).

Some Kalman filtering approaches could also be considered for use as gating functions. The dynamic process model can be given by mechanistic model equations, and a nonparametric model can function as the measurement model, which is used to iteratively update/correct the predictions of the process model. The weighting of the knowledge sources is accomplished by the tuning parameters of the filter (covariance matrices of the measurement noise and system disturbances). Simutis and Lübbert have applied an unscented Kalman filter, consisting of a dynamic material balance model and support vector machines–based measurement model, to an *E. coli* fermentation process (Simutis and Lübbert 2017) (see Chapter 5).

2.2.4 Static/Dynamic Models

Whether to develop a static or a dynamic model will depend on the objective of the model development exercise. Static models will suffice if distribution, monitoring, or similar problems need to be solved. Dynamic models are required when the temporal behavior of a system is of interest—for instance, to understand the interactions or manipulate (control) some of the system states.

Though the objectives are different, in principle the development of static hybrid models will not differ from dynamic ones, and the structures shown in Figure 2.1 represent both static and dynamic approaches.

The static model can be described by

$$Y = h(X, p) \tag{2.5}$$

where:
$h(\cdot)$ represents a hybrid modeling function
p the set of parameters contained in both the parametric and nonparametric model

Similarly, the dynamic model can be represented by

$$\frac{dY}{dt} = h(X, Y, p) \tag{2.6}$$

There exist two structural differences between the static and dynamic model: (1) dynamic models need an additional numerical integration step to obtain the state; and (2) the dynamic model predictions can be fed back to the model (Y can appear in the hybrid modeling function), which gives rise to auto-dynamic behavior, such as oscillations, and so on (Seborg et al. 2016).

For the identification of the parameter values in both modeling approaches, for example, the following least-squares objective function can be minimized:

$$J = \frac{1}{\sigma} \cdot (Y_{mes} - Y)^2 + \lambda \cdot \|p\|_m \qquad (2.7)$$

where:
Y_{mes} are the measured outputs
σ represents a set of weighting factors (e.g., the variance of Y)
λ is a set of regularization factors
m describes which norm of the parameters is chosen, typically 1, 2, or 0

The second term (the regularization term) can be added to drive the values of those parameters that are not significantly contributing to the model fit to zero, promoting lean modeling structures. The minimization of the objective function is typically accomplished with gradient-based approaches. In the case of the static model, it is straightforward to obtain the analytical gradients, dY/dp; for the dynamic model, the so-called sensitivities approach is used (Psichogios and Ungar 1992; Oliveira 2004). For the identification of the parameters of the nonparametric model and discrimination of the structure, the same strategies described earlier for nonparametric models are adopted.

2.2.5 Hybrid Modeling Structures for Process Operation and Design

Key to a successful application of hybrid models for process monitoring, control, optimization, or scale-up is a considerate model structure that reflects the particular requirements of the application. Therefore, the developers of hybrid models should bear in mind the modeling objective or the particular application: What is the model developed for?

For process monitoring, the focus is on the actual state of the process at the moment of inquiry and its comparison to the optimal and/or past trajectories. A static model structure will allow prediction of the state of the process at the moment of inquiry, based on on-line measurements of related variables. With the same measurements, a dynamic model might predict the state at the next time step, predicting one step ahead. Multi-step-ahead predictions of the state can be made if the measurements can be replaced by their expected values—if the inputs to the model do not comprise dependent variables. Structures that allow multi-step-ahead predicting thus only rely on independent variables (control degrees of freedom) and/or their own

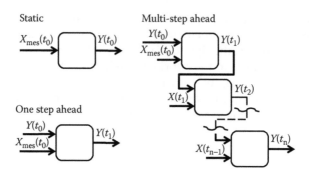

FIGURE 2.3
Parallel hybrid model structure.

prediction of the state for prediction of the next step (see Figure 2.3). For instance, consider the monitoring of concentration c_A in a reactor operated in continuous mode using a spectroscopic device. The measured intensities, I, correlate to c_A; however, the correlation might depend on the temperature, T. One could come up with the static hybrid model $c_A = f(T) \cdot I$, assuming that the correlation coefficient is a function of T. Another approach would be to use the dynamic material balances, assuming that the content of the reactor is ideally mixed (homogenous):

$$\frac{dc_A}{dt} = r(T, I) - D \cdot (c_A - c_{A,f})$$

where:
   D is the dilution rate
   $c_{A,f}$ is the concentration of compound A in the inlet
   $r(T, I)$ is the reaction rate

This rate is modeled via a data-driven approach and uses T and I as inputs, thus using an actual measure of c_A for the predictions. The intensities will depend on c_A and thus are not known a priori, for which reason one can only predict one time step ahead. However, the model could be formulated differently, yielding a multi-step-ahead prediction structure:

$$\frac{dc_A}{dt} = r(T, c_A) - D \cdot (c_A - c_{A,f})$$

This structure does not need the intensities to predict the concentration levels for future times. The originally proposed static model could now be used to correct the dynamic model at certain time instances—for example, using a Kalman filtering approach.

Process control requires an understanding of the dynamic, and thus dynamic modeling structures are preferential. Whether one- or multi-step-ahead predictor structures are to be preferred depends on the choice of control framework. Optimal control, model predictive control, and like approaches require the predictions for a certain time horizon, necessitating multi-step-ahead predictor structures. Model reference control, generic model control, or similar approaches also work with one-step-ahead predicting structures, since these approaches only use the prediction of the state (and its dynamics) for the moment at which the control action is computed. However, in both cases the underlying models should describe the dynamics in the region where the process is normally operated and in regions beyond normal operation, such that the controller can get the process back on track even for large disturbances.

Consider, for instance, the monitoring example with the aim to control the concentration c_A tightly around a specific set point because the reaction r is very exothermic. The dynamics of the concentration c_A is sought to be controlled by changing the feeding concentration $c_{A,f}$. The control action is computed and utilizes the dynamics predicted by the model, $c_{A,f} = m(dc_A/dt, K)$, with K the controller parameters and $m(\cdot)$ the controller function.

In the case that c_A moves away from regions on which $r(c_A, T)$ was developed, the data-driven model cannot be expected to predict reliably. Mis-predictions in $r(c_A, T)$ will result in mis-predictions in the dynamics (dc_A/dt), and thus the manipulation of $c_{A,f}$ under consideration of the modeled dynamics could potentially result in a disaster. If, however, the reaction can be modeled by $r(T) \cdot c_A$, then the prediction of the data-driven model is independent of c_A and thus any deviation in c_A does not result in model mis-predictions. Consequently, good control can be achieved even if c_A and the dynamics as well become very low or great. This property of hybrid models is referred to as frequency extrapolation (Van Can et al. 1998), in an analogy to the frequency domain in the field of process control.

Process optimization typically seeks to explore a user-defined space for the optimal process conditions. Depending on the optimization objective (dynamic versus steady-state optimization), dynamic and static model structures can be used for this purpose. Ideally one would like to encounter the process optimum without characterizing the entire search space experimentally. Thus, the frequency and also range extrapolation properties of hybrid models (Van Can et al. 1998) are of particular importance for process optimization. Range extrapolation means that the model inherits a good prediction performance beyond the range of values tested experimentally. Consider the reformulated process control example:

$$\frac{dc_A}{dt} = r(T) \cdot c_A - D \cdot (c_A - c_{A,f})$$

which has two degrees of freedom, T and $c_{A,f}$. There is an interplay between the inlet concentration and the cooling requirements/costs. For instance, increasing $c_{A,f}$ will result in an increase in c_A and thus in an increase in heat released by the exothermic reaction. The cooling of the reactor needs to account for this increase to keep the temperature constant and operation safe. Depending on the price of the resulting reaction product, the price of $c_{A,f}$, and the cost of cooling, the optimal operation conditions will vary. Due to the provided explicit reaction relation ($r(T) \cdot c_A$), operation at different c_A levels can be considered without losing model validity as long as the temperature can be maintained within safe operation regions (i.e., the cooling requirements are lower than cooling capacity). This optimization could even be accomplished without consideration of the dynamics ($dc_A/dt = 0$), showing that a static hybrid model is sufficient for optimization of steady-state operations.

Scale-up of processes can be difficult because the underlying phenomena that govern the process behavior might change, which requires an experimental reevaluation of the process behavior at the new scale. While it is difficult to capture or describe these changes in the underlying phenomena, models can help evaluate the changes in the time scales and/or with designing the dimensions of the equipment. However, to have a chance of describing the process at the larger scale, the model should be formulated such that the predicted quantities are independent of the dimensions. Taking, for instance, the concentration example, the modeled concentration dynamics are independent of the volume of the reactor (and of the operation mode—i.e., batch, fed-batch, or continuous). If instead the dynamics of the mass of compound A would have been described

$$\frac{dm_A}{dt} = \frac{dV \cdot c_A}{dt} = V \cdot r(T) \cdot c_A - u \cdot c_A + u \cdot c_{A,f}$$

with u the inlet feeding rate, the model would only be valid for the dimension on which it had been developed. The property to extrapolate in an extra dimension that was not explicitly studied during model development is referred to as dimension extrapolation (Van Can et al. 1998). Te Braake et al. (1998) provide some rules on how to develop models that can be used with/for scale-up.

Summarizing, the choices are

1. whether the structure should be static or dynamic
2. whether, in case of a dynamic structure, one- or multi-step-ahead properties are required
3. which part of the model is to be represented by the data-driven approach and which part could be modeled explicitly

2.3 Examples

2.3.1 Example 1: Static Parallel Hybrid Model of a Ribosome Binding Site in Bacteria

2.3.1.1 Problem Statement

Quantitative sequence-activity modeling (QSAM) is essential for engineering synthetic DNA sequences with desired biologic activity. The problem may be defined as the prediction of protein titer or related biological activity as a function of the nucleotide sequence of a certain DNA/RNA sequence that regulates its expression. Due to the combinatorial nature of nucleotide sequences, exploring the whole design space by random mutation is experimentally impractical. In order to decrease the experimental effort, rational design methods supported by QSAM can be employed. The keystone of QSAM is the ability to predict the biologic activity of novel DNA/RNA sequences. Such predictive power has been proven difficult due to the length of design sequences, resulting in complex highly dimensional designs that suffer from the *curse of dimensionality* problem. In this example, we study hybrid semiparametric QSAM in the context of data sparsity, using the 5′UTR region in *Escherichia coli* as an illustrative example. The objective of the QSAM model is to predict the protein expression level as a function of the nucleotide sequence of the 5′UTR in *E. coli* with a maximum length of 164 bp. The basic assumption is that initiation is the limiting step in the protein translation mechanism (Salis et al. 2009). This allows the reduction of the target design sequence to 75 bp around the protein start codon.

2.3.1.2 Data Set

The data set published by Espah Borujeni et al. (2013) was adopted for model development, where the 5′UTR sequences were designed and inserted into six different bacteria, including two *E. coli* strains, expressing four different reporter proteins. The data set comprises 485 mRNA sequences, each with a varying size between 19 and 164 bp, and respective protein fluorescence levels. For more details, see Espah Borujeni and Salis (2016). The data were divided into a model identification partition and a test partition. The model identification partition served to identify model structure and to estimate the underlying parameter values. The test partition served to assess the model predictive power. More specifically, data partitions were randomly selected from the uniform distribution, with 67% of sequences for model identification and 33% of sequences for model testing. The procedure was repeated 100 times, yielding 100 different models, to eliminate a possible data sampling effect.

2.3.1.3 Hybrid Model Structure

The hybrid model structure is represented schematically in Figure 2.4a. It consists of a parallel hybrid structure composed of two modules. The first module is the thermodynamic model (TM) proposed by Salis et al. (2009). The protein expression is modeled as a function of the total free Gibbs energy of variation (ΔG_{TOT}) for the formation of ribosome-mRNA complex:

$$\Delta G_{TOT} = \Delta G_{mRNA:rRIB} + \Delta G_{START} + \Delta G_{SPACING} - \Delta G_{STANDBY} - \Delta G_{mRNA} \quad (2.8)$$

which accounts for five free Gibbs energy terms corresponding to the different steps for translation initiation (for details see Espah Borujeni and Salis 2016). Then the protein titer is given by

$$P_{TM} = \alpha\, t\, e^{-\beta \Delta G_{TOT}} \quad (2.9)$$

with α, a calibration parameter that is specific for a given expression system that accounts for translation-independent parameters such as the DNA copy

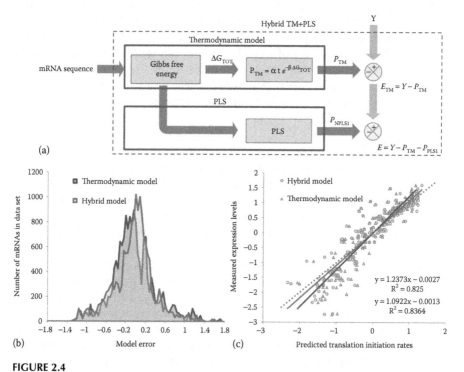

(a)

(b)

(c)

FIGURE 2.4
Comparison of modeling results of the standalone TM model and the hybrid TM+PLS model: (a) parallel hybrid model structure; (b) prediction residuals distribution; and (c) predicted *versus* measured protein titer.

number, the promoter transcription rate and the mRNA stability, t the cultivation time, and β the Boltzmann factor.

The second module is a projection to latent structures (PLS) model that runs in tandem with the TM and extracts information from the TM residuals as a function of mRNA primary structure. The inputs to the PLS are the five individual free Gibbs energy terms used to calculate ΔG_{TOT} through Equation 2.9 and also the ΔG_{TOT}. The PLS target output consists of the TM model residuals. Thus the hybrid model decomposes the target (measured) protein vector $\mathbf{Y}$ in 3 terms:

$$\mathbf{Y} = \mathbf{P}_{TM} + \mathbf{P}_{PLS} + \mathbf{E} \qquad (2.10)$$

The first two terms represent the contribution of the TM and PLS to the prediction of $\mathbf{Y}$. The vector $\mathbf{E}$ is the final hybrid model residuals.

2.3.1.4 Hybrid Model Identification

The model identification was performed in two steps as follows:

Step 1: TM identification. First, the TM model was fitted by linear regression to the model identification data set. Then the TM residuals, $\mathbf{E}_{TM}$, were calculated:

$$\mathbf{E}_{TM} = \mathbf{Y} - \mathbf{P}_{TM} \qquad (2.11)$$

Step 2: PLS identification. The N-PLS MATLAB® implementation described in Andersson and Bro (2000) was used. The PLS had six inputs and one target output. The optimal number of latent variables was determined by the Akaike's Information Criterion with second-order bias correction (AICc) method (Burnham and Anderson 2004).

Finally, it should be noted that all terms of Equation 2.10 were normalized in the same way—by applying the natural logarithm and then by auto-scaling.

2.3.1.5 Results and Discussion

Table 2.1 shows the mean squared error (MSE) for calibration (model identification data partition) and for prediction (test data partition) for the standalone TM model, the standalone PLS model, and the hybrid TM+PLS model. The TM model shows only a slightly higher prediction MSE than the calibration MSE. This is a good indication that the TM model is not overfitting the calibration data set, which is not surprising as the TM model contains only two parameters. Moreover, the fact that the prediction error is

TABLE 2.1

Mean Squared Error Obtained for the Thermodynamic (TM) Model, PLS Model, and Hybrid Parallel TM+PLS Model

	Calibration	Prediction
TM	0.19	0.20
PLS	0.24	0.27
Hybrid TM+PLS	0.17	0.18
Ratio Hybrid/TM	0.89	0.90

only slightly higher than the calibration error indicates that the TM model is already a highly predictive model.

Comparing the standalone PLS and TM models, it can be concluded that both the calibration and prediction errors increase in relation to the standalone PLS despite the fact that the exact same information is used in both models. The number of latent variables selected by the AICc method is only two, which represents in this case a good trade-off between calibration error and prediction error. Increasing the number of latent variables significantly reduces the calibration MSE (eventually below that of the TM model), but at the cost of a much higher prediction error. Thus it may be concluded that the TM model structure is a clear advantage in relation to a purely empirical method such as the PLS.

As for the hybrid TM+PLS model, the results show that both the calibration and the prediction MSE are decreased in relation to the standalone TM by 11% and 10%, respectively, which is a significant improvement. This improvement is achieved merely by the definition of a more suitable model structure, given that the data set and input/output variables are all the same for both models. Figure 2.4b compares the TM and hybrid model residuals distribution. Figure 2.4c compares the hybrid and TM model protein titer predictions against the respective measurements. The mean of the residuals for both models is ~0. However, the standard deviation of the residuals is reduced by 7% (0.44 for the TM down to 0.41 for the hybrid TM+PLS).

2.3.2 Example 2: A Dynamic Bootstrap Aggregated Hybrid Modeling Approach for Bioprocess Modeling

Data play an important part in the development of hybrid models: (1) they are used to discriminate the structure of the nonparametric parts; and (2) they are used to identify the values of the parameters, which can be contained in both the parametric and nonparametric parts. In many cases the number of data points is low. This makes the choice of which part of the data to use in the training, validation, or test partition particularly difficult, since this choice might have a significant impact on the performance of the model for novel conditions (unseen data).

The common strategy to overcome this limitation is to (1) repartition the data; (2) develop one model on every new partition; and (3) aggregate the model, after a certain number of models have been developed—to compute the weighted average of their response. This strategy, referred to as bootstrap aggregating (bagging), was proposed in 1994 (Breiman 1996), and it has been applied to artificial neural networks (Zhang 1999), decision trees (Prasad et al. 2006; Svetnik et al. 2003), PLS models (Mevik et al. 2004; Carinhas et al. 2011), and more.

In the field of hybrid modeling, this strategy has so far received only limited attention. Zhang and coworkers used bootstrap aggregated neural networks in hybrid models and found their performance to be superior to that of standard neural networks (Tian et al. 2001). Carinhas et al. applied this strategy to calculate the confidence limits for a PLS model, which was used in a hybrid model (Carinhas et al. 2011). Peres et al. developed a method that identifies several nonparametric models in parallel and assigns a weighting to their contribution as an input to the parametric part of the structure (Peres et al. 2001). However, to date no bootstrap aggregated hybrid model has been developed.

2.3.2.1 Methodology

2.3.2.1.1 Bootstrap Aggregated Hybrid Models

Relative to the number of available experiments, the data are partitioned into different sets (e.g., for three experiments, three different separations into training and validation partitions are possible); however, a maximum number of sets can be specified. On each of these sets, a hybrid model is developed following the approach described earlier: (1) the data are partitioned into three sets—training, validation, and test partitions; and (2) the training partition is used for parameter (weight) identification (using the sensitivities approach), and the validation partition is used to decide when to stop the training. However, the test partition is used only for the aggregated hybrid model, to discriminate between the optimal nonparametric model structures; it is assumed that regardless of the data, the structure of the nonparametric model is identical. Not all of the models that were developed on the different sets contribute to the bootstrap aggregated model, but only the best ~30% performing models are combined into the aggregated hybrid model (i.e., models with unsatisfactory performance are not included, since they would decrease the overall performance and would counteract the original idea of *robustifying* the model performance against variations in the data). The aggregation is performed by averaging the predictions obtained for each variable at each time point. The standard deviation can also be computed from the predictions, such that *confidence* intervals are obtained for the aggregated predictions.

2.3.2.1.2 *Simulation Case Study:* E. coli *Process Model*

A simulation case study is used to demonstrate the potential of the boot-strap aggregated hybrid model. A fed-batch *E. coli* process for the production of a viral capsid protein is simulated, comprising the concentration of biomass, substrate, and the protein. The specific growth, uptake, and product formation kinetics are functions of temperature and substrate concentration. The control degrees of freedom (factors) are the substrate feeding rate and temperature. A Doehlert design was used to investigate the impact of temperature (three levels) and the feeding rate (five levels) on the process response, which resulted in the execution (simulation) of nine experiments. The simulated process data were corrupted with 5% white noise. The model equations can be found in von Stosch and Willis (2016).

The structure of the hybrid model is based on the material balances of biomass and protein (product):

$$\frac{dX}{dt} = \mu \cdot X - D \cdot X \tag{2.12}$$

$$\frac{dP}{dt} = v_P \cdot X - D \cdot P \tag{2.13}$$

where:
   X and P are the biomass and product concentrations, respectively
   D is the dilution rate
   μ and v_P are the specific biomass growth and product formation rates, respectively

The rates (whose functions are typically unknown) are here modeled using an artificial neural network. The inputs of the artificial neural network are the biomass concentration, the feeding rate, and the temperature. The neural network contains three layers, the nodes of the input and output layer have linear transfer functions, and the nodes of the hidden layer have tangential hyperbolic transfer functions. Three nodes in the hidden layer proved to be sufficient to model the underlying function (captured in the data).

2.3.2.2 *Results and Discussion*

In Figure 2.5, the fit of the model predictions to the experimental data is shown. It can be seen that the model excellently fits the data for both concentrations. The standard deviations for the prediction of the product

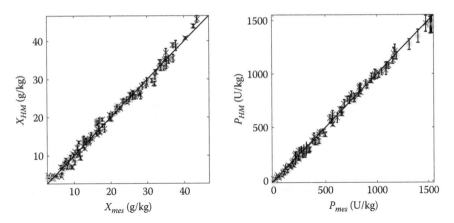

FIGURE 2.5
Prediction of the bootstrap aggregated hybrid model versus experimental data for biomass and product concentrations. Dark gray x's represent predictions for the training and validation set; light gray stars represent predictions for the test set; black vertical bars represent standard deviations of predictions.

concentrations are slightly larger than those for the biomass concentrations, in particular for the test data. This is somewhat surprising, as the test set contains an experiment that is a repetition experiment at the center point. However, the counterpart center-point experiments contained in the training-validation set have fewer points at higher product concentrations; therefore, this region is not well represented during the model development. This shows the need for relatively frequent and continuous sampling. It can be seen in Figure 2.6 that the bootstrap aggregated hybrid model describes the shape of the product response surface at the end process time much more accurately than does the hybrid model developed on a single set of data. Though this is not surprising, it emphasizes that an aggregated hybrid model with better prediction performance can be obtained with the same amount of data. In addition, standard deviations are obtained for the predictions. These give an indication to which degree the predictions can be *trusted* and thus help make better development or optimization decisions.

It is shown that the aggregated hybrid models outperform a single hybrid model, which is in agreement with the findings by Zhang et al. (Tian et al. 2001) for bootstrap aggregated neural networks. In addition, the standard deviations of the predictions can be computed, which provides a confidence measure for assessing the quality of the predictions. The bootstrap aggregated hybrid models can be expected to be of particular interest for development and optimization studies and in cases where only a low amount of data is available.

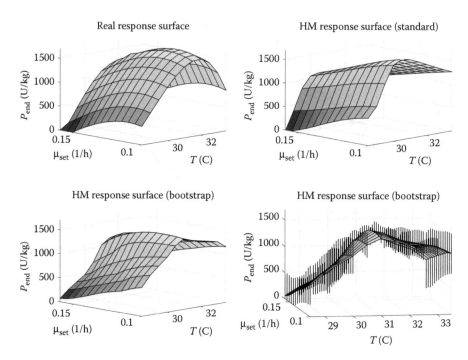

FIGURE 2.6
Product concentrations at tf = 17 h, over different process conditions computed for the process simulation, a hybrid model developed in the standard way (using a single data set), and the bootstrap aggregated hybrid model with and without standard deviations.

2.4 Concluding Remarks

In many engineering problems, a mechanistic modeling framework is too costly for practical implementation. This is typically the case in multi-scale modeling problems, where macroscopic observations are modeled as a function of molecular-scale material properties. Also, product quality attributes that are determined by molecular-scale properties are particularly difficult to model. With the increasing availability of high-throughput analytical techniques, empirical nonparametric modeling methods are often preferred over mechanistic parametric methods. The point highlighted in this chapter is that the two sources of knowledge and modeling methods are not mutually exclusive. When both sources of process knowledge are insufficient individually, hybrid models may offer an opportunity to build a sufficiently accurate model. In many published studies, hybrid models have been reported to outperform their standalone components (a topic covered in the remaining chapters of this book). However, this should not be taken as a universal rule. Some of the key hybrid structures and respective identification methods were reviewed and illustrated with two case studies. The keystone for successful hybrid model identification can be summarized as the ability to find the optimal combination between mechanistic parametric structure and empirical nonparametric structure, which obviously depends on the knowledge resources available.

References

Andersson, C. A. and R. Bro. 2000. The N-way toolbox for MATLAB. *Chemometrics and Intelligent Laboratory Systems* 52 (1): 1–4. doi: 10.1016/S0169-7439(00)00071-X.

Breiman, L. 1996. Bagging predictors. *Machine Learning* 24 (2): 123–140. doi: 10.1007/bf00058655.

Burnham, K. P. and D. R. Anderson. 2004. Multimodel inference. *Sociological Methods and Research* 33 (2): 261–304. doi: 10.1177/0049124104268644.

Carinhas, N., V. Bernal, A. P. Teixeira, M. J. T. Carrondo, P. M. Alves, and R. Oliveira. 2011. Hybrid metabolic flux analysis: Combining stoichiometric and statistical constraints to model the formation of complex recombinant products. *BMC Systems Biology* 5 (1): 34. doi: 10.1186/1752-0509-5-34.

Engle, R. F., C. W. J. Granger, J. Rice, and A. Weiss. 1986. Semiparametric estimates of the relation between weather and electricity sales. *Journal of the American Statistical Association* 81 (394): 310–320. doi: 10.1080/01621459.1986.10478274

Espah Borujeni, A., A. S. Channarasappa, and H. M. Salis. 2013. Translation rate is controlled by coupled trade-offs between site accessibility, selective RNA unfolding and sliding at upstream standby sites. *Nucleic Acids Research* 42 (4): 2646–2659. doi: 10.1093/nar/gkt1139.

Espah Borujeni, A. and H. M. Salis. 2016. Translation initiation is controlled by RNA folding kinetics via a ribosome drafting mechanism. *Journal of the American Chemical Society* 138 (22): 7016–7023. doi: 10.1021/jacs.6b01453.

Ichimura, H. 1993. Semiparametric least squares (SLS) and weighted SLS estimation of single-index models. *Journal of Econometrics* 58 (1): 71–120. doi: 10.1016/0304-4076(93)90114-K.

Kramer, M. A. and F. Eric Finch. 1988. Development and classification of expert systems for chemical process fault diagnosis. *Robotics and Computer-Integrated Manufacturing* 4 (3): 437–446. doi: 10.1016/0736-5845(88)90015-4.

Mevik, B.-H., V. H. Segtnan, and T. Næs. 2004. Ensemble methods and partial least squares regression. *Journal of Chemometrics* 18 (11): 498–507. doi: 10.1002/cem.895.

Nair, A. and R. Daryapurkar. 2016. Analysis of fluidized catalytic cracking process using fuzzy logic system. *2016 International Conference on Advances in Computing, Communications and Informatics (ICACCI)*. Jaipur, India: IEEE, September 21–24, 2016.

Nascimento Lima, N. M., F. Manenti, R. M. Filho, M. Embiruçu, and M. R. Wolf Maciel. 2009. Fuzzy model-based predictive hybrid control of polymerization processes. *Industrial and Engineering Chemistry Research* 48 (18): 8542–8550. doi: 10.1021/ie900352d.

Oliveira, R. 2004. Combining first principles modelling and artificial neural networks: A general framework. *Computers and Chemical Engineering* 28 (5): 755–766. doi: 10.1016/j.compchemeng.2004.02.014.

Peres, J., R. Oliveira, and S. Feyo de Azevedo. 2001. Knowledge based modular networks for process modelling and control. *Computers and Chemical Engineering* 25 (4–6): 783–791. doi: 10.1016/S0098-1354(01)00665-2.

Peres, J., R. Oliveira, and S. Feyo de Azevedo. 2008. Bioprocess hybrid parametric/nonparametric modelling based on the concept of mixture of experts. *Biochemical Engineering Journal* 39 (1): 190–206. doi: 10.1016/j.bej.2007.09.003.

Prasad, A. M., L. R. Iverson, and A. Liaw. 2006. Newer classification and regression tree techniques: Bagging and random forests for ecological prediction. *Ecosystems* 9 (2): 181–199. doi: 10.1007/s10021-005-0054-1.

Psichogios, D. C. and L. H. Ungar. 1992. A hybrid neural network-first principles approach to process modeling. *AIChE Journal* 38 (10): 1499–1511. doi: 10.1002/aic.690381003.

Qian, Y., X. Li, Y. Jiang, and Y. Wen. 2003. An expert system for real-time fault diagnosis of complex chemical processes. *Expert Systems with Applications* 24 (4): 425–432. doi: 10.1016/S0957-4174(02)00190-2.

Salis, H. M., E. A. Mirsky, and C. A. Voigt. 2009. Automated design of synthetic ribosome binding sites to control protein expression. *Nature Biotechnology* 27 (10): 946–950.

Seborg, D. E., T. F. Edgar, D. A. Mellichamp, and F. J. Doyle. 2016. *Process Dynamics and Control*, 4th ed. Hoboken, NJ: Wiley Global Education.

Simutis, R. and A. Lübbert. 2017. Hybrid approach to state estimation for bioprocess control. *Bioengineering* 4 (1). doi: 10.3390/bioengineering4010021.

Simutis, R., R. Oliveira, M. Manikowski, S. Feyo de Azevedo, and A. Lübbert. 1997. How to increase the performance of models for process optimization and control. *Journal of Biotechnology* 59 (1–2): 73–89. doi: 10.1016/S0168-1656(97)00166-1.

Svetnik, V., A. Liaw, C. Tong, J. C. Culberson, R. P. Sheridan, and B. P. Feuston. 2003. Random forest: A classification and regression tool for compound classification and QSAR modeling. *Journal of Chemical Information and Computer Sciences* 43 (6): 1947–1958. doi: 10.1021/ci034160g.

Te Braake, H. A. B., H. J. L. Van Can, and H. B. Verbruggen. 1998. Semi-mechanistic modeling of chemical processes with neural networks. *Engineering Applications of Artificial Intelligence* 11 (4): 507–515. doi: 10.1016/S0952-1976(98)00011-6.

Teixeira, A. P., J. J. Clemente, A. E. Cunha, M. J. T. Carrondo, and R. Oliveira. 2006. Bioprocess iterative batch-to-batch optimization based on hybrid parametric/nonparametric models. *Biotechnology Progress* 22 (1): 247–258. doi: 10.1021/bp0502328.

Thompson, M. L. and M. A. Kramer. 1994. Modeling chemical processes using prior knowledge and neural networks. *AIChE Journal* 40 (8): 1328–1340. doi: 10.1002/aic.690400806.

Tian, Y., J. Zhang, and J. Morris. 2001. Modeling and optimal control of a batch polymerization reactor using a hybrid stacked recurrent neural network model. *Industrial and Engineering Chemistry Research* 40 (21): 4525–4535. doi: 10.1021/ie0010565.

Van Can, H. J. L., H. A. B. Te Braake, S. Dubbelman, C. Hellinga, K. C. H A. M. Luyben, and J. J. Heijnen. 1998. Understanding and applying the extrapolation properties of serial gray-box models. *AIChE Journal* 44 (5): 1071–1089. doi: 10.1002/aic.690440507.

von Stosch, M. and M. J. Willis. 2016. Intensified design of experiments for upstream bioreactors. *Engineering in Life Sciences*. doi: 10.1002/elsc.201600037.

Willis, M. J. and M. von Stosch. 2016. Inference of chemical reaction networks using mixed integer linear programming. *Computers and Chemical Engineering* 90: 31–43. doi: 10.1016/j.compchemeng.2016.04.019.

Zhang, J. 1999. Developing robust non-linear models through bootstrap aggregated neural networks. *Neurocomputing* 25 (1–3): 93–113. doi: 10.1016/ S0925-2312(99)00054-5.

3

Hybrid Models and Experimental Design

Moritz von Stosch

CONTENTS

3.1 Introduction

The development of models from data that were collected during normal operation of the process is a challenging task, if not impossible. This is because there exists only a little variation in the data, which makes the inference of cause-and-effect relationships from the data difficult. The integration of process knowledge, along with data-driven models, into so-called hybrid models can, to some degree, compensate for poor data quality (Fiedler and Schuppert 2008). However, the quality of the data (the root problem for inference) can be improved by designing the experiments in such way that cause-and-effect relationships can be inferred (i.e., using DoE methods). This means that the data should contain systematic variations in those variables that are inputs to the data-driven model, such that their responses in the output variables can be discerned. Typically, DoEs comprise several experiments, and as a consequence their application is rather scarce at manufacturing scale but common in process development at lab/pilot scale.

The DoE concept has been applied in several areas (Montgomery 2008) and has recently attracted a lot of interest in the (bio)pharmaceutical field due to novel guidelines (Mandenius and Brundin 2008; Kumar et al. 2014). The concept has also found application to support the development of hybrid models (von Stosch et al. 2014); however, it is not clear which DoEs are the most suitable for the development of hybrid models. Also, in the case of hybrid models in particular, the integration of dependent variables and a priori knowledge about the process should be considered. Both points will be addressed in this chapter.

Model development is typically only the first step, and the models are usually exploited for process operation or design (monitoring, control, optimization, or scale-up). It is of particular importance for process optimization and scale-up to assess the reliability of the model's predicted process conditions. The DoE and the ranges of the investigated process parameters characterize a process operation region/domain, which in turn determines the validity/applicability region of the model. In particular, the nonparametric model can only be expected to predict reliably close to the region of input values on which it was developed. Therefore, the consideration of the validity domain is very important for process optimization and scale-up, where the optimal process conditions are sought, which might not fall within the previously investigated process region. Different measures are available that allow the investigation of the validity/applicability of the hybrid model (Leonard et al. 1992; Teixeira et al. 2006; Kahrs and Marquardt 2007; von Stosch et al. 2014) and will be discussed in this chapter.

The systematic extension of the validity domain and/or the systematic improvement of the hybrid model via the addition of data can principally be accomplished in two ways, both involving the model. Either the model is used to optimize the process, or the model suggests how the input domain of the nonparametric model can be better spanned by maximizing some measure of difference between the historical and new process conditions. After the experiments have been carried out, the data that become available are used to improve the process model. The model is then eventually used to further improve the process/model. In this chapter, it will be shown that both strategies can also be combined, which is particularly interesting in light of the ever-increasing amount of parallel experimentation that is becoming available.

3.2 Design of Experiments

3.2.1 Basics of Design of Experiments

Design of experiment (DoE) approaches are used to plan systematic changes in the system's parameters/variables (referred to as factors), such that their impact on the response of the system can, to some degree, be distinguished. Two principal types of factors can be differentiated: controllable and uncontrollable factors (Montgomery 2008). Controllable factors can be varied such that their impact on the response of the system can systematically be evaluated. The variation in the system's response that is due to uncontrollable factors can be evaluated with techniques such as replication, randomization, and blocking.

Changes in the controllable factors can result in different responses of the system, describing a linear or nonlinear behavior of the system. Different levels of the factors are typically evaluated to account for nonlinearities; these include, for example, evaluation of temperature (factor) at three levels—25°C, 30°C, and 35°C. At least three levels are required to observe nonlinear behavior. Also, interactions between factors can occur: the system's response for simultaneous changes in two factors is different from the sum of their responses changing one factor at a time. Thus, simultaneous variations in the factors need to be evaluated to understand the interactions, which makes additional experiments necessary.

3.2.2 Various Design of Experiment Strategies

The planning and type of DoEs differs with the objective of the study and the knowledge available about the system. Three objectives can be differentiated: (1) scoping and screening, (2) development and optimization, and (3) analysis of robustness and variance.

For scoping and screening designs, typically only the main effects are evaluated, eventually in combination with interlevels, to assess the degree of nonlinearity. These designs are used to locate a particular region of interest and to evaluate which factors should be included in a more detailed analysis/design. An example is the fractional factorial design.

Development and optimization designs are employed to identify the process optimum and the associated value of each factor. These designs are typically comprised of fewer factors than screening designs, and also a much lower range of values is studied for each factor. These designs are based on the idea that the solution surface can be approximated by a quadratic function, which typically only holds true in close vicinity of the optimum. The reduction of each factor's studied range and the containment of the optimum in this range is therefore of major importance for the success of these designs. Examples of these types of designs are central composite, Box-Behnken or Doehlert designs, which are typically used along with a quadratic model also referred to as a response surface method.

Another DoE objective can be to make the system insensitive to variation in noise factors, which is achieved by using robust parameter designs such as Taguchi designs.

3.2.3 Design of Experiments for Hybrid Models

At the beginning of the model development, the inputs of the nonparametric model are typically unknown (i.e., nonparametric models with different inputs will be studied during the model development). Moreover, some of the inputs of the nonparametric model might not be varied independently— for example, when using estimated quantities in dynamic hybrid models as inputs (feedback).

Due to these issues, it is not clear a priori which DoE strategy will benefit the hybrid model development the most. Therefore, the focus is on studying the impact of the (main) factors on the system's response. Multiple designs have been applied, along with hybrid models, for this purpose. Factorial designs have frequently been used with hybrid models (Tholudur and Ramirez 1999; Gupta et al. 1999; Tholudur et al. 2000; Thibault et al. 2000; Saraceno et al. 2010). Chang et al. used a uniform design to develop the model and then used a sequential pseudo-uniform design to extend the validity region, which is similar to an optimal experimental design procedure (Chang et al. 2007). In this contribution, the process optimization performance obtained for different DoEs, listed in Table 3.1, is compared.

A limitation of the classic DoEs is their rather static nature: the level of the factors is kept constant throughout the experiment. Therefore, the impact of dynamic variations in the factors on the process might not be captured. While this is typically not critical for process development and optimization, the response to dynamic variations is important for process control. One possibility is to incorporate dynamic variations into the hybrid model by performing intra-experiment changes of the factors according to a chosen DoE. This approach was recently investigated and is referred to as intensified design of experiments (iDoE) (von Stosch et al. 2016b). iDoEs allow the development of a model that can be used for process optimization and control. The iDoE strategy is obtained on the basis of a classic DoE. A specified number of experiments of the classic DoE are grouped together in one experiment, and the different levels of the factors are evaluated during each dynamic experiment.

TABLE 3.1

Design of Experiments

DoE Method	Type of Design	Levels for T, X_{ind}, and μ_{Set}	No. of Experiments
Main effect design + threefold center-point repetition	Screening design	(2, 2, 2) + 3 × center point	9
Doehlert design	Space-filling optimization design	(7, 3, 5) + 3 × center point	15
Inscribed central composite design	Optimization design	(5, 5, 5) + 3 × center point	17
Box-Behnken design	Optimization design	(3, 3, 3) + 3 × center point	15
iDoE based on two 2-factor (T, μ_{Set}) Doehlert designs: (1) X_{ind} = 8 (g/kg) and (2) X_{ind} = 15 (g/kg)	Optimization design	2 × ((3, 5) + 3 × center point)	10

In order to account for process stage–dependent behavior, the grouped DoEs are run through twice, but during different process phases: for example, upon going through a series of points in direction 1, 2, 3, and 4, they are then run counter-rotating—4, 3, 2, and 1 (see also Figure 3.1). The analysis of iDoEs cannot be accomplished with traditional response surface methods but requires the application of models that take into account the variant nature. In order to compare the iDoE to the other designs, it is also included in the analysis.

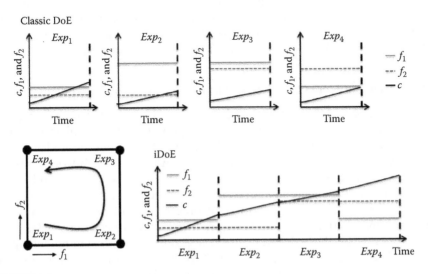

FIGURE 3.1
A schematic representation of the classic DoE and the iDoE approaches.

3.2.4 *E. coli* Case Study: Design of Experiments for Hybrid Model Development

3.2.4.1 Design of Experiments

In practical cases, the factors, the range of each factor, and the type of design have to be chosen. All of these choices should be aim/objective oriented. In other words, which factors have (or are likely to have) an impact on the objective, in which range can the optimum of the objective be expected, and which design is the most adequate to reach the objective (see Section 3.2.2)? The *E. coli* case study focuses on the maximization of the final product titer in an upstream bioprocess—optimization of the fermentation step. The process parameters that are known to impact on the final product titer are temperature, pH, carbon feeding rate, and induction time or biomass concentration at induction, among others. In this study we focus on temperature (T), the biomass concentration at induction (X_{ind}), and the desired specific biomass growth rate (μ_{Set}) as a replacement for the carbon feeding rate (see Appendix 3.A for details). The ranges have been chosen based on typically experienced values: 0.1–0.16 (1/h), 29.5–33.5 (C), and 5–19 (g/l), respectively. The designs listed in Table 3.1 were implemented for these factors and ranges and the experiments simulated with the equations shown in Appendix 3.A.

It can be assumed a priori that the material balances might provide a valid modeling framework (see also Appendix 3.B). In these balances, the reaction terms are typically represented by data-driven models. The reaction rates usually depend on the temperature, pH, substrate concentrations, inhibitor concentrations, and so on. Whereas temperature and pH can be chosen independently, many of the other concentrations depend on the behavior of the system and can thus not be readily excited using the DoE methodology. However, variations in the feeding rate and initial concentrations might provide sufficient excitation to discriminate the model. The approaches introduced in Section 3.4 can then be used to improve the model further.

3.2.4.2 Hybrid Model Development from the Designs of Experiments

The conditions of the different DoEs were used to generate data using the simulated fed-batch *E. coli* case, shown in Appendix 3.A. A number of hybrid models were developed for each DoE, varying the number of nodes in the hidden layer of the artificial neural network (ANN). For each DoE, the best-performing hybrid model was determined as described in Appendix 3.B. The hybrid models were then used to predict the product titer at the end of the fermentation ($P_{HM}(t_f)$) for different process conditions; in other words, the response surfaces for the final product titer were calculated. The difference between the *true* response surface (obtained with the model equations of the simulation case, Appendix 3.A) and the predictions of the hybrid models are shown for each DoE in Figure 3.2. The *true* response surface is also shown. It can be observed that the greatest product titers are obtained

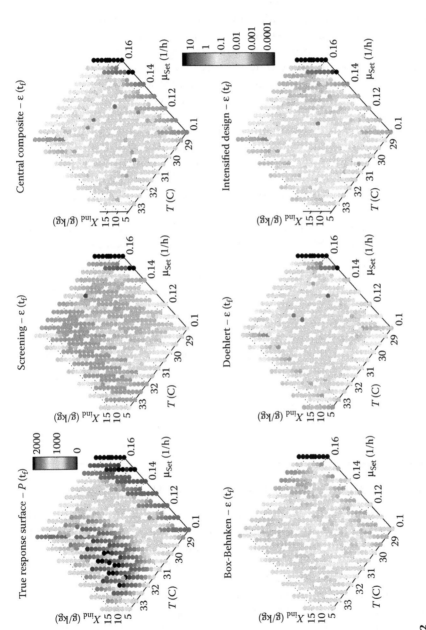

FIGURE 3.2

The *true* response surface and the error between the *true surface* and the predictions of the hybrid models developed on the data obtained for different designs over temperature, feeding rate, and initial biomass concentration.

for high initial biomass concentrations, a temperature of about 32°C, and a desired specific growth rate of 0.12 (1/h). The greatest prediction errors are obtained at 28.5°C and desired specific growth rate of 0.16 (1/h), irrespective of the initial biomass concentration and also irrespective of the DoE. Among all DoEs, the screening DoE seems to provide the worst approximation of the true system, as the prediction errors are the greatest over the studied region. These errors decrease only in direct vicinity to the experimentally tested conditions, which were used to develop the model. The models developed upon the data obtained for the Doehlert and central composite designs seem to provide the best description across the spanned region. In the case of the central composite design, this is not too astonishing, as it involves the greatest number of experiments, scattered across the entire region. However, in case of the Doehlert design, the number of experiments is lower, but the uniform distribution of the experiments across the spanned region seems to help cover the systems behavior better. However, the extreme points/ends of the ranges are better described in the case of the Doehlert than in that of the central composite design. In the case of the Box-Behnken design, which comprises the same number of experiments as the Doehlert design, the description of the overall region is worse; however, the limits of the spanned region are described better here than in the case of the Doehlert and central composite designs. In the case of the intensified design, which just comprises one experiment more than the screening design, the obtained descriptions seem better than for the Box-Behnken and the screening designs and only slightly inferior to the central composite and Doehlert designs. These observations agree well with the findings of other studies; see, for example, von Stosch et al. (2016a). It seems that it is important to characterize the space as uniformly as possible in order to yield a model that provides a good process description across the searched process region.

3.3 The Validity/Applicability Domain of Hybrid Models

The validity/applicability domain of hybrid models is affected by (1) the incorporated fundamental (parametric) knowledge, (2) the experimental data, and (3) the input domain of the nonparametric model.

Typically, the limitations for the incorporated parametric knowledge are known; for example, kinetic models might only be valid close to equilibrium or where there is an unlimited availability of co-reactants, and thus can be explicitly accounted for by the modeler. The input domain of the nonparametric models is usually a priori not entirely known and thus determined during the model development process. The inputs domain is generated from the functional dependencies of the outputs on certain input variables. It might be known that an output depends on certain variables (though not necessarily how, since this is to be determined using the nonparametric model), but additional dependencies might be discovered. For example, it is generally known that the substrate uptake rate depends on the substrate concentration, but it could also be discovered that an inhibitor addition- ally affects it. In an ideal case, the variation in every input variable is such that their impact on the output can be distinguished from that of the other inputs. To ensure this, each input would ideally be included as a factor in a DoE. However, in practical cases there arise two problems, as already out- lined earlier: (1) not all input dependencies might be a priori known, and (2) the input variables might vary throughout an experiment and cannot be independently chosen. Therefore, the DoE is planned in such a way as to understand the impact of the factors of interest on the system's response, as outlined earlier. During the development of the nonparametric model, dif- ferent topologies are investigated, which includes testing different inputs and different structures of the nonparametric model. The nonparametric model can only be expected to accurately predict the response of the system for combinations of input values on which it has been trained and validated. Thus, measures are required that characterize the validity domain of the nonparametric model.

3.3.1 Approaches to Characterize the Validity/Applicability Domain

Different measures have been reported to characterize the validity/ applicability domain (Teixeira et al. 2006; Kahrs and Marquardt 2007; von Stosch et al. 2014). Three measures have found application with hybrid mod- els and are discussed in more detail: a convex hull, clustering, and a confi- dence interval approach. The former two judge upon the likely prediction quality of the model based on the position of a novel input value combina- tion with respect to the training input combinations. The latter provides a confidence measure for the prediction of the hybrid model.

3.3.1.1 Convex Hull

The convex hull of a set of data points is the smallest convex set that con- tains all of the points. It can be computed from the input values of the training data using, for example, the Qhull algorithm (Barber et al. 1996). Upon the characterization of the convex hull, the position of new data with respect to the convex hull can be calculated using a set of linear equations (Kahrs and Marquardt 2007). Thus, this criterion can be implemented as a set of linear inequality constraints for process optimization, which is com- putationally efficient and reduces the search space to the domain to within the convex hull.[*]

---

[*] Linear constraints are more computationally efficient to handle than are nonlinear constraints.

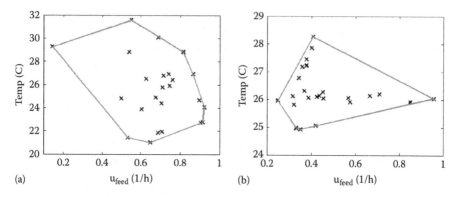

FIGURE 3.3
Simulated data and the respective convex hull for two demonstrative cases: (a) far-out point (outlier) and (b) missing interactions.

However, this criterion might be too optimistic in case of a nonuniform distribution of the input data. Two extreme cases are shown in Figure 3.3. In case A, the position of the far-out data point extends the validity domain significantly, but the system response in the region without data points has not been evaluated and might differ from the nonparametric model estimations. In case B, the data points gather along the two axes, and for that reason the model will mostly capture the main effects, but the convex hull also includes regions of interactions between the variable on axes 1 and 2. This can be critical, since the interaction between variables can alter the system's response (which is, for instance, explicitly accounted for in DoEs).

3.3.1.2 Clustering

The characterization of the input domain using a clustering approach has been described by Teixeira et al. (2006) and Leonard et al. (1992). The clustering has typically been accomplished using the K-means algorithm. This algorithm randomly distributes k cluster centers across the input domain and then refines their position by minimizing the distance to n_X nearest data points (this number can be chosen by the user). The width of every cluster is then derived from the average distance of the cluster center to the nearest data points. For new data points, their position in relation to the nearest cluster can be evaluated, and, based on the distance of the new data point to the characterized input domain, a degree of membership is obtained. The degree of membership is a value between 0 and 1. It can be used to define a measure of risk, which is obtained by using $r = 1 - m$.

This criterion generally handles nonuniform data distributions better than does the convex hull criterion. However, if the data are sparse or if the number of nearest data points is chosen too large, then the clusters become too wide and thus the criterion too optimistic.

For the optimization, a desired level of maximum risk can be chosen. The maximum risk level is typically achieved using the validation and test data. This maximum risk level is then used to either constrain the local risk r or the cumulative risk. One possible cumulative risk measure is the time average risk:

$$R = \frac{1}{t_f - t_0} \cdot \int_{t_0}^{t_f} (1-m) \cdot dt \qquad (3.1)$$

The risk constraint is nonlinear, which computationally is more expensive than the linear convex hull constraint.

3.3.1.3 Confidence Interval

The convex hull and clustering criterion both focus on the input domain. In contrast, the estimation of confidence intervals provides a measure that considers the data distribution and the influence of measurement noise on the model prediction. Kahrs and Marquardt (2007) proposed using a maximum likelihood approach for the confidence interval estimation since it provides a good balance between computational requirements and performance (see also Papadopoulos et al. 2001). They proposed minimizing the prediction error

$$\min_{p}\{E\} = \det((c_{mes} - c(t,p))^T \cdot (c_{mes} - c(t,p))) \qquad (3.2)$$

where:
   c_{mes} is a vector of measured concentration values
   $c(t,p)$ are the model estimates
   p are the model parameters

This error criterion is identical to the least-squares estimate under the assumption of normal-Gaussian distributed errors with unknown variance. The confidence interval can then be expressed as $c(t,p) \pm q \cdot s$, where q can either be obtained assuming a t-distribution, $q = t(n_D - n_p, 1 - \alpha/2)$ (Leonard et al. 1992), or an F-distribution, $q = F(n_p, n_D - n_p, 1 - \alpha)$ (Kahrs and Marquardt 2007), with $n_D - n_p$ degrees of freedom (n_D number of data and n_p number of parameters) and significance level α and where:

$$s = \sigma \cdot \sqrt{2 \cdot g^T \cdot H^{-1} \cdot g} \qquad (3.3)$$

$$\sigma = \sqrt{E/(n_D - n_p)} \qquad (3.4)$$

In these equations, $g = dc(t,p)/dp$ and $H = d^2E/dp^2$ (which at the optimum can be approximated by $H = dE^T/dp \cdot dE/dp$). The F-distribution is the more exact measure for multivariate parameter distributions. Other methods for the estimation of the confidence intervals can also be applied—for example, bootstrapping, which is particularly powerful for nonlinear models but computationally more expensive. The confidence interval is used with a qualitative measure that helps assess the model predictions.

3.3.2 *E. coli* Case Study: The Validity Domain and Relation to the Design of Experiments

The DoE can have a direct impact on the validity domain, in particular if the factors are inputs to the nonparametric model. The systematic exploration of the process operation domain will in principle also help yield a model that can describe the process for the entire characterized process operation region.

3.3.2.1 *Clustering and Convex Hull Approaches*

An example of the validity domain, characterized by clustering and convex hull approaches, is shown in Figure 3.4 along with the data used to characterize the domain. For the data stem from the experiments simulated according to the Doehlert design, see Table 3.1. The different levels in temperature can be clearly recognized. However, the biomass concentration at induction and also the changes in the desired specific growth rate (which are translated into changes of the feeding rate) are much harder to see. Nevertheless, it is evident that the systematic variations in these control degrees of freedom span the validity domain; for instance, the three experiments at 32°C start in the same range but then diverge from one another, because of different levels in the feeding rate. The data are inside of the convex hull, and some of the process trajectories constitute the corners of the hull. In the case of the clustering approach (the borders calculated for a local risk of 20%), the separate clusters can still be identified and the characterized validity domain is larger than in the case of the convex hull approach. However, a gap in the domain can be seen between temperatures of around 31°C and 34°C for low biomass concentrations and low feeding rate values. This gap arises from the different operating conditions, which have separated the process trajectories. With an increasing risk level, this gap closes. In the case of the convex hull, a similar gap does not arise, since the conditions are covered within the convex region. Although the gap in the present case is not critical (as the process conditions are fairly close), this observation makes clear the difference between the clustering and the convex hull approach.

3.3.2.2 *Confidence Interval*

The confidence interval approach can be carried out in addition to the clustering/convex hull approach. In Figure 3.5, selected data obtained for

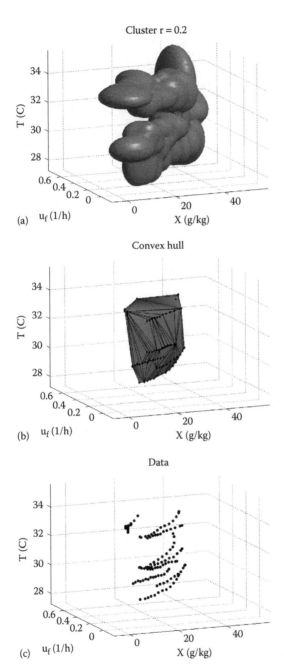

FIGURE 3.4
Validity domain for the nonparametric model input data of the hybrid model developed
on the basis of the Doehlert design: (a) clustering, (b) convex-hull, and (c) data used for the
characterization.

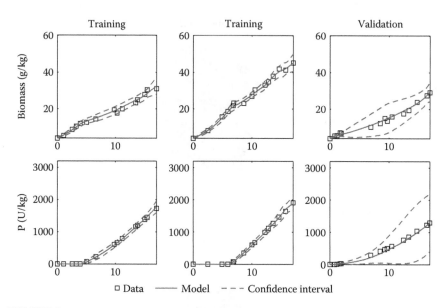

FIGURE 3.5
Biomass and product concentrations over time for selected experiments of the screening design. Black squares: experimental data; continuous line: hybrid model predictions; dashed lines: 95%-confidence interval limits.

the screening experiment are shown together with the model estimations/predictions and the confidence intervals. Also, the predictions and confidence intervals for final product titer maximization (see Section 3.3.1.3) are shown. The confidence band is very narrow for the (shown) training experiments, widening slightly from the beginning of the process to the end. For the validation experiment, the confidence band is considerably wider for the biomass concentration and even wider for the product concentration, where, in the case of the product, the band again widens toward the end. The shown validation experiment was carried out at considerably different process conditions—lowest biomass induction concentration, temperature, and desired growth rate—which explains the wide confidence interval. However, the agreement between the model estimations and the data is excellent for the training and validation sets.

3.4 Hybrid Model-Based (Optimal) Experimental Design

The validity domain of the hybrid models can be systematically extended. Though the validity domain is predominantly determined by the experimental data, the systematic extension of the validity domain can be accomplished regardless of which data have been used to develop the hybrid model. Two fundamental approaches for the extension of the validity domain can be differentiated: (1) iterative batch-to-batch optimization and (2) model-based experimental design. The objective of the latter approach is to span the nonparametric model's input domain in such a way that it is coherently extended even for dependent variables. The objective of the former approach is the directed extension of the validity domain toward the region that covers the process optimum. These objectives can also be combined, as will be addressed.

3.4.1 Iterative Batch-to-Batch Optimization

The iterative batch-to-batch optimization is an approach for identifying the process region in which optimal process performance can be yielded. This approach has successfully been applied by Teixeira et al. (2006) and Ferreira et al. (2013). A schematic representation is shown in Figure 3.6. The initial model is developed based on an excitation data set, which should include variations in the factors of interest. Consequently, the model is used to predict the conditions under which optimal process performance is obtained—for example, maximizing end-point product titer. The maximization is subject to the hybrid model equations, and the risk and physical (e.g., maximum working volume), chemical, and/or biological (e.g., enzyme temperature sensitivity) constraints (see Ferreira et al. 2013).

Then the predicted performance and process parameters are compared to previously established experiments, and in the case that significant changes in performance and parameters exist, the optimization experiment is executed. Experimentally obtained performance is compared to predicted performance. The data of the optimization experiments are combined with the previous data, and the model is redeveloped based on all data. The cycle is then followed through several times until no further increase in process performance is predicted.

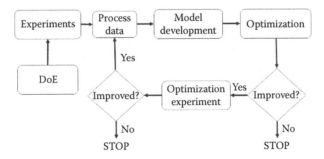

FIGURE 3.6
A schematic representation of the iterative batch-to-batch approach.

3.4.2 Model-Based Experimental Design

Model-based experimental design—also referred to as optimal experimental design—is commonly used to reduce the uncertainty of parameter values or to discriminate between competing models. In neural network research, a similar approach exists, typically referred to as active learning, although the idea there is to better explore the region in which the model predictions can be assumed to be of poor quality rather than to directly improve the structure or parameters (weights) of the model. For instance, the sequential pseudo-uniform design approach, which was proposed by Chang et al. (2007) and used along with a hybrid model on a simulated polymerization reactor system, seeks to explore the design region more efficiently. Brendel et al. (2008) proposed a similar approach that, in particular, seeks to extend the design region of dependent variables. Thus, in contrast to classical DoEs, the factors in their approach are not required to be independent, as long as the model describes the dependencies. Brendel and colleagues show that the approach is more efficient in exploring a process region than are classical DoEs. The incorporation of the dependencies is a particular strength of this approach, since the model will capture the process dynamics—those arising from the feedback/dependencies—much better. This is of particular importance for process control, since feedback/dependencies affect the system response and stability. The approach can be set up as an optimization problem, where the difference between the factor values of the new experiment and those of the already-executed experiments is maximized (Brendel and Marquardt 2008).

3.4.3 Process Optimum Directed Model-Based Experimental Design

High-throughput experimentation is getting increasingly popular in the process industries. Instead of doing a number of experiments in series, high throughput allows one to do a certain number of experiments at a time. In light of this trend, the series approach taken in the iterative batch-to-batch schema is inefficient, but the combination of the iterative with the optimal experimental design approach can be expected to efficiently explore the process region. Therefore, it is proposed to perform the optimization and the directed exploration of the solution space at the same time. This can be established by optimizing the process conditions first, followed by an exploration of the process operation region near the predicted optimal conditions. In this way, the region of interest is explored in a more directed and efficient manner.

3.4.4 *E. coli* Case Study

3.4.4.1 *Iterative Batch-to-Batch*

The objective function in the *E. coli* case study is the maximization of the end-point product titer:

$$\max_u \{P_{HM}(t_f)\} \qquad (3.5)$$

The control degrees of freedom $u = [u_f, T, X_{ind}]^T$ are the feeding rate, temperature, and biomass concentration at the point of induction. The limits of the control degrees of freedom are assumed to be the same as for the DoE, with the exception of the biomass concentration at induction, for which the range of 5–40 g/kg was chosen. The clustering approach is used to constrain the optimization to regions in which the model is assumed to produce accurate predictions; therefore, the time-averaged risk obtained by Equation 3.1 has to be below a chosen maximum risk level. For each model, the maximum risk level was chosen separately, since it depends on the training and validation data. The optimization is further subject to Equations 3.B1 and 3.B2 (see chapter appendix). The risk levels and optimization results for each of the DoEs are specified in Table 3.2. Approximately the same number of experiments was required to encounter the optimal process region, despite the different number of experiments used in the DoEs. This result is in line with the previous experience of the author. A possible explanation is that the iterative

TABLE 3.2

Results Obtained for the Iterative Batch-to-Batch and Process Optimum–Directed Model-Based Experimental Design

Design	Experiments[a]	Nodes[b]	Performance: Predicted/Experimental[c]	Risk
Screening	6 + 2 + 1	4	3156/1993	15.7%
	7 + 2 + 1 (+ 1)	4	2139/2147	5.9%
Box-Behnken	8 + 4 + 3	5	2101/1779	14.3%
	9 + 4 + 3	5	2084/2113	22.4%
	10 + 4 + 3	7	2123/1821	10.6%
	11 + 4 + 3 (+ 1)	7	2053/2248	8.0%
Doehlert	8 + 4 + 3	5	2350/845	25.3%
	10 + 4 + 3	5	2468/2126	19.6%
	11 + 4 + 3 (+ 1)	7	2292/2233	12.0%
Central composite	9 + 4 + 4	5	2583/2051	18.9%
	10 + 4 + 4	5	2275/2169	13.2%
	11 + 5 + 3 (+ 1)	6	2224/2222	12.2%
Screening + parallel	6 + 2 + 1	4	3156/1993, (2263/2034, 2009/886, 2137/1571)	15.7%, (2.8%, 8.0%, 5.8%)
	8 + 3 + 2 (+ 4)	5	2116/2260, (2076/2196, 2116/1977, 2146/2234)	15.43%, (7.5%, 11.8%, 13.1%)
iDoE	6 + 2 + 2	4	1908/1727	13.3%
	7 + 2 + 2	5	2496/1800	7.5%
	7 + 3 + 2	7	2425/2303	21.2%

[a] The separations in the experiments column refer to Train + Valid + Test + (optimization experiment).
[b] Number of nodes in the hidden layer of the ANN.
[c] The square brackets in this column designate the additional parallel experiments.

batch-to-batch optimization represents an active learning approach that tries to achieve a better model accuracy in the region of interest—the region of the process optimum. However, the success of this strategy depends critically on the quality of the excitation data set and the region that is covered by these data. The reason the quality of the excitation data is so critical is that even though better extrapolation properties have been reported for hybrid models, those factors, which enter into the nonparametric model, will limit the extrapolation properties (accounted for by the validity measure).

The final optimization result (the final product titer) varies slightly from DoE to DoE (Table 3.2). In each case, the optimization was stopped, as there exist only minor differences between the optimal process conditions, which in reality would not be explored further. This can also be seen in Figure 3.7, which shows the iteration paths of the process conditions obtained with the models that had been developed on data obtained from different DoEs. Astonishingly, the hybrid model developed on the data obtained for the screening design required the fewest iterations and number of experiments. This result is not representative, however, and the final titer is also the lowest obtained for all models. Nevertheless,

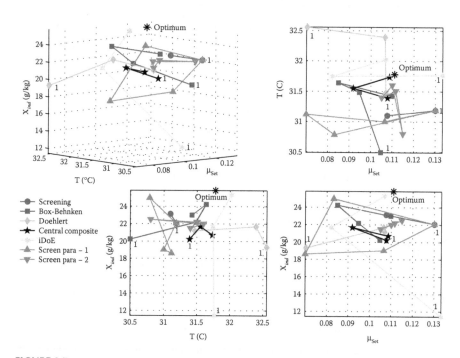

FIGURE 3.7
The optimal process conditions predicted during the batch-to-batch iterations for the hybrid models trained on data obtained from different DoEs.

the continual application of the hybrid model together with a screening design many times provides a good basis for a directed search of the process optimum. The results obtained for the iDoE seem very promising. Although three iterative steps were required to approach the optimal process region, this approach required only 13 experiments in total to get to the optimal region. This approach also yielded the greatest final titer, getting the closest to the true optimum. These results also make sense in light of the excellent model performance across the region spanned by the DoE, which can be seen in Figure 3.2.

3.4.4.2 Process Optimum–Directed Model-Based Experimental Design

In step 1, the final product titer is maximized, identical to the procedure described in Section 3.4.4.1. In a second step, the experimental conditions for a fixed number of parallel experiments, n_{para}, is optimized:

$$\max_{u} \left\{ \prod_{i=1}^{n_{para}} \prod_{j=i+1}^{n_{para}} \prod_{k=1}^{n_u} \left| \frac{u_{k,i} - u_{k,j}}{s_k} \right| \right\} \tag{3.6}$$

where:
n_u is the number of the degrees of freedom
s_k are scaling coefficients
$u = [u_F, T, X_{ind}]^T$

The multiplications ensure that the conditions are different in all degrees of freedom. Sums could be used instead, relaxing this constraint and making it similar to the function proposed by Brendel and Marquardt (2008). The optimization is subject to the model in Equations 3.B1–3.B2, the constraints for the degrees of freedom, the risk constraint (Equation 3.1), and an additional constraint,

$$P_{HM}(t_f) > P_{HM,opt} \cdot S_{opt} \tag{3.7}$$

which ensures that the n_{para} experiments are in an area that exhibits a final titer that is S_{opt} times the one of the optimal titer (in this case $S_{opt} = 0.9$).

As in the case of the screening design, this method required two iterations to approach the optimum (Table 3.2). Figure 3.7 demonstrates that in the first iteration the area in which a close-to-optimal titer is obtained is relatively large. In the second iteration, the predicted optimal process region is much smaller and relatively close to the true optimum, indicating that the approximation of the optimal process region in the second iteration is much better than in the first iteration and also better than for the single screening approach. However, the final titer achieved is greater than for the screening design–based iterative approach, which is most likely due to the fact that the region near the process optimum is better explored.

3.5 Conclusions

Hybrid models combine parametric and nonparametric models. The development of the nonparametric model requires data that contain sufficient excitation to determine the impact of the nonparametric model inputs on its outputs. The necessary excitations can be achieved by design of experiment approaches, though special attention should be given to the dependent nonparametric model inputs. Several different DoEs—Box-Behnken, Doehlert, central composite, screening, and iDoE designs—have been implemented, and hybrid models were developed on the obtained data. It was observed that the best model predictions for the region covered by the DoE were obtained for the models developed on data of the central composite design, the Doehlert design, and the iDoE, because the experiments of these DoEs are spread across the covered region in a more homogeneous way. The latter two designs involve significantly less experiments than the central composite design. The iDoE approach seems particularly interesting.

Approaches for the characterization of the validity domain of the model predictions—such as convex hull, clustering, and confidence interval approaches—were compared next. The convex hull and clustering approaches focus on the input domain of the nonparametric model, since the nonparametric can be expected to predict reliably only in the vicinity of the conditions used for the development of the model. The clustering approach characterizes the validity domain better than the convex hull approach in the case of nonuniformly distributed data. In addition, the confidence interval approach provides a direct measure for the confidence in the predicted values—the model outputs.

In the last section, approaches for the systematic extension of the validity domain were discussed that (1) aimed to explore the region of the process optimum better (iterative batch-to-batch approaches), (2) sought to extend the nonparametric model input region in a better way (optimal experimental design approaches), or (3) combined both previously named approaches. It was shown that regardless of the DoE used to develop the model, the iterative batch-to-batch approach required only a few experiments to get close to the process optimum. In the case of the combination of the iterative batch-to-batch with the optimal experimental design procedure, it was found that the process region in which the optimum is encountered is explored in a better way. The combination of the approaches is becoming particularly interesting in light of high-throughput experimentation.

Appendix

3.A Simulation Case

The production of viral capsid protein production by *E. coli* was simulated using the model proposed by Levisauskas et al. to exemplify the discussed approaches. The model comprises the material balances for biomass (X), substrate (S), and product concentration (P) and the overall mass balance (W):

$$\frac{dX}{dt} = \mu \cdot X - D \cdot X \tag{3.A1}$$

$$\frac{dS}{dt} = -v_S \cdot X - D \cdot (S - S_F) \tag{3.A2}$$

$$\frac{dP}{dt} = v_P \cdot X - D \cdot P \tag{3.A3}$$

$$\frac{dW}{dt} = u_F \tag{3.A4}$$

with μ, v_S, and v_P the specific rates of biomass growth, substrate uptake, and product formation; $D = u_F/W$ the dilution rate; and u_F the feeding rate.

The specific biomass growth rate was modeled using the expression

$$\mu = \mu_{max} \cdot \frac{S}{S + K_S} \cdot \frac{K_i}{S + K_i} \cdot \exp(\alpha \cdot (T - T_{ref})) \tag{3.A5}$$

where:
$\mu_{max} = 0.737$ (1/h)
$K_S = 0.00333$ (g/kg)
$K_i = 93.8$ (g/kg)
$\alpha = 0.0495$ (1/C)
$T_{ref} = 37$ (C)
T is the temperature of the experiment

The specific substrate uptake rate is modeled via

$$v_S = \frac{1}{Y_{XS}} \cdot \mu + m \tag{3.A6}$$

with $Y_{X/S} = 0.46$ (g/g) and $m = 0.0242$ (g/g/h).
The specific product formation rate is

$$v_P = \frac{I_D}{T_{PX}} \cdot \left(\frac{p_{X,max} \cdot \mu \cdot k_m}{k_\mu + \mu + \mu^2/k_{i\mu}} - p_X \right) \tag{3.A7}$$

with

$$p_{X,max} = \frac{A_e \cdot \exp\left(\dfrac{-H_1}{R \cdot (T + 273.15)}\right)}{\left(1 + B_e \cdot \exp\left(\dfrac{-H_2}{R \cdot (T + 273.15)}\right)\right)} \tag{3.A8}$$

where:

$T_{PX} = 1.495$ (h)
$k_\mu = 0.61$ (1/h)
$k_m = 751$ (U/g)
$k_{i\mu} = 0.0174$ (1/h)
$A_e = 5 \cdot 10^5$
$K_e = 3 \cdot 10^{93}$
$H_1 = 62$
$H_2 = 551$
$R = 8.31446 \cdot 10^{-3}$

the induction parameter $I_D = 0$ before induction and $I_D = 1$ afterwards.

For the feeding rate, an exponential profile was adopted to match a desired constant specific biomass growth, μ_{Set}, as follows:

$$u_F = \frac{1}{S_F \cdot Y_{XS}} \cdot \mu_{Set} \cdot X_0 \cdot W_0 \cdot \exp(\mu_{Set} \cdot (t - t_0)) \tag{3.A9}$$

where:

$X_0 = X(t_0)$ (g/kg) is the initial biomass concentration
$W_0 = W(t_0)$ (kg) is the initial weight of the culture broth

The process was divided into two phases—a growth and a production phase. During the growth phase, $\mu_{Set} = 0.5$ (1/h) and $T = 27$ (C). The duration of the growth phase t_I is implicitly subject to investigation by the design of experiment, which includes $X_I = X(t_I)$ (g/kg), the biomass concentration at induction, as follows:

$$t_I = \frac{1}{\mu_{Set}} \cdot \ln\left(\frac{S_F \cdot X_I - X_I \cdot X_0 \cdot 1/Y_{XS}}{S_F \cdot X_0 - X_I \cdot X_0 \cdot 1/Y_{XS}}\right) \tag{3.A10}$$

For the production phase, the levels of μ_{Set} and T also were investigated, using the design of experiment.

3.B The Hybrid Model

The material balances provide a sound and generally valid modeling framework that has commonly been used as the backbone for several hybrid models. The balance equations for biomass and product (with X_{HM} and P_{HM} designating biomass and product concentrations, respectively) are written using specific rates:

$$\frac{dX_{HM}}{dt} = \mu_{ANN} \cdot X_{HM} - D \cdot X_{HM} \qquad (3.B1)$$

$$\frac{dP_{HM}}{dt} = v_{P,ANN} \cdot X_{HM} - D \cdot P_{HM} \qquad (3.B2)$$

with $D = u_F/W$, the dilution; μ_{ANN}, the specific biomass growth rate; and $v_{P,ANN}$ the specific product formation rate. The two specific rates are modeled using an artificial neural network. The network has three layers (input, hidden, and output layer), which typically is sufficient for the modeling of arbitrary continuous nonlinear functions. The transfer functions of the nodes of the layers are linear, tangential hyperbolic, and linear, respectively. Three inputs were identified to be sufficient to approximate the rate functions—those of the temperature, the feeding rate, and the model estimated biomass concentration. The performance of different numbers of nodes in the hidden layer of the neural network was studied. The neural network that performed best in terms of the BIC was chosen for the optimization studies. The BIC balances the fit of the model estimates and experimental data against the number of model parameters (i.e., model complexity). For instance, in the case of equal performance in terms of fit (MSE), the model with the lower number of parameters is chosen.

For the training, validation, and testing of the model performance, the available data were separated into three corresponding partitions: training, validation, and test partitions. The parameters were adapted to minimize a weighted least-square function of the concentrations using the training data. The validation data were used to stop the training once the fit of the model estimates and the experimental data for the validation data did not further improve. The training was 40 times reinitiated from random parameter values, and the best-performing parameter set was chosen, thus avoiding local minima. The test partition was used to evaluate the model performance on new, unseen data. For more details, see Oliveira (2004) and Teixeira et al. (2006).

References

Barber, C. B., P. D. David, and H. Hannu. 1996. The quickhull algorithm for convex hulls. *ACM Transactions on Mathematical Software* 22 (4): 469–483. doi: 10.1145/235815.235821.

Brendel, M. and W. Marquardt. 2008. Experimental design for the identification of hybrid reaction models from transient data. *Chemical Engineering Journal* 141 (1–3): 264–277. doi: 10.1016/j.cej.2007.12.027.

Chang, J.-S., S.-C. Lu, and Y.-L. Chiu. 2007. Dynamic modeling of batch polymerization reactors via the hybrid neural-network rate-function approach. *Chemical Engineering Journal* 130 (1): 19–28. doi: 10.1016/j.cej.2006.11.011.

Ferreira, A. R., J. M. L. Dias, M. Stosch, J. Clemente, A. E. Cunha, and R. Oliveira. 2013. Fast development of Pichia pastoris GS115 Mut+ cultures employing batch-to-batch control and hybrid semi-parametric modeling. *Bioprocess and Biosystems Engineering* 37 (4): 629–639. doi: 10.1007/s00449-013-1029-9.

Fiedler, B. and A. Schuppert. 2008. Local identification of scalar hybrid models with tree structure. *IMA Journal of Applied Mathematics* 73 (3): 449–476. doi: 10.1093/imamat/hxn011.

Gupta, S., P.-H. Liu, S. A. Svoronos, R. Sharma, N. A. Abdel-Khalek, Y. Cheng, and H. El-Shall. 1999. Hybrid first-principles/neural networks model for column flotation. *AIChE Journal* 45 (3): 557–566. doi: 10.1002/aic.690450312.

Kahrs, O. and W. Marquardt. 2007. The validity domain of hybrid models and its application in process optimization. *Chemical Engineering and Processing: Process Intensification* 46 (11): 1054–1066. doi: 10.1016/j.cep.2007.02.031.

Kumar, V., A. Bhalla, and A. S. Rathore. 2014. Design of experiments applications in bioprocessing: Concepts and approach. *Biotechnology Progress* 30 (1): 86–99. doi: 10.1002/btpr.1821.

Leonard, J. A., M. A. Kramer, and L. H. Ungar. 1992. A neural network architecture that computes its own reliability. *Computers and Chemical Engineering* 16 (9): 819–835. doi: 10.1016/0098-1354(92)80035-8.

Levisauskas, D., V. Galvanauskas, S. Henrich, K. Wilhelm, N. Volk, and A. Lübbert 2003. Model-based optimization of viral capsid protein production in fed-batch culture of recombinant *Escherichia coli*. *Bioprocess and Biosystems Engineering* 25 (4): 255–262. doi: 10.1007/s00449-002-0305-x.

Mandenius, C.-F. and A. Brundin. 2008. Bioprocess optimization using design-of-experiments methodology. *Biotechnology Progress* 24 (6): 1191–1203. doi: 10.1002/btpr.67.

Montgomery, D. C. 2008. *Design and Analysis of Experiments*, 7th ed. New York: John Wiley & Sons.

Oliveira, R. 2004. Combining first principles modelling and artificial neural networks: A general framework. *Computers and Chemical Engineering* 28 (5): 755–766.

Papadopoulos, G., P. J. Edwards, and A. F. Murray. 2001. Confidence estimation methods for neural networks: A practical comparison. *IEEE Transactions on Neural Networks* 12 (6): 1278–1287. doi: 10.1109/72.963764.

Saraceno, A., S. Curcio, V. Calabrò, and G. Iorio. 2010. A hybrid neural approach to model batch fermentation of "ricotta cheese whey" to ethanol. *Computers and Chemical Engineering* 34 (10): 1590–1596. doi: 10.1016/j.compchemeng.2009.11.010.

Teixeira, A. P., J. J. Clemente, A. E. Cunha, M. J. T. Carrondo, and R. Oliveira. 2006. Bioprocess iterative batch-to-batch optimization based on hybrid parametric/nonparametric models. *Biotechnology Progress* 22 (1): 247–258.

Thibault, J., G. Acuña, R. Pérez-Correa, H. Jorquera, P. Molin, and E. Agosin. 2000. A hybrid representation approach for modelling complex dynamic bioprocesses. *Bioprocess Engineering* 22 (6): 547–556. doi: 10.1007/s004499900110.

Tholudur, A. and W. F. Ramirez. 1999. Neural-network modeling and optimization of induced foreign protein production. *AIChE Journal* 45 (8): 1660–1670. doi: 10.1002/aic.690450806.

Tholudur, A., W. F. Ramirez, and J. D. McMillan. 2000. Interpolated parameter functions for neural network models. *Computers and Chemical Engineering* 24 (11): 2545–2553. doi: 10.1016/S0098-1354(00)00615-3.

von Stosch, M., J.-M. Hamelink, and R. Oliveira. 2016a. Hybrid modeling as a QbD/ PAT tool in process development: An industrial *E. coli* case study. *Bioprocess and Biosystems Engineering* 39 (5): 773–784. doi: 10.1007/s00449-016-1557-1.

von Stosch, M., J.-M. Hamelink, and R. Oliveira. 2016b. Towards intensifying design of experiments in upstream bioprocess development: An industrial *E. coli* feasibility study. *Biotechnology Progress*. doi: 10.1002/btpr.2295.

von Stosch, M., R. Oliveira, J. Peres, and S. Feyo de Azevedo. 2014. Hybrid semi-parametric modeling in process systems engineering: Past, present and future. *Computers and Chemical Engineering* 60 (0): 86–101. doi: 10.1016/j. compchemeng.2013.08.008.

4

Hybrid Model Identification and Discrimination with Practical Examples from the Chemical Industry

Andreas Schuppert and Thomas Mrziglod

CONTENTS

4.1 Introduction

Data-based modeling is well established as a powerful approach for modeling of the input-output behavior of complex systems in various application areas. It is the method of choice if the mechanism of the system is not fully understood and the efforts to close the gaps in our knowledge are not affordable. Numerous realizations and the respective numerical implementations are available. Evidence in application projects, however, shows that many applications suffer significantly from problems that are characteristic of data-based modeling: the curse of dimensionality and the lack of extrapolability. Moreover, in application projects it is not satisfactory to exclude the available knowledge, even if it is not complete. A systematic integration of incomplete knowledge and data-based modeling would be more favorable. Hence, hybrid modeling aims to provide a flexible approach to adapt modeling to the availability of data and knowledge on the one side and the requirements on predictive power on the other side. In this chapter, some mathematical background is first outlined that is critical to the applications presented in the remainder of the chapter.

4.2 Why Data-Based Modeling?

Many applications in chemical engineering, biotechnology, and medicine aim to control or optimize extremely complex systems that are not yet fully understood. Therefore, the only choice available to researchers is to either ignore these applications because they look to be intractable or to develop and provide techniques that allow them to span bridges over the limited knowledge. In an abstract manner, the problem can be formulated as follows:

There is a complex system for which a set of n input parameters can be observed (and partially controlled):

$$x = \{x_1, \ldots, x_n\} \in \mathfrak{R}^n \, z \in \mathfrak{R}^1$$

The system produces an output that can be characterized by one or more observable output values z. The output variables z can only be observed and in no case directly controlled. However, the output variables z can be assumed to depend, in an appropriate approximation, on the observable input variables x. Hence, the existence of a system function F mapping each input value x within an admissible range Ω onto a unique value $z(x)$ can also be assumed:

$$z = F(x) + O(\varepsilon), \, \varepsilon \ll 1$$

For a wide range of applications, the output z of the complex system is to be analyzed, controlled, or optimized using the controllable subset of the input parameters x in order to achieve the required output z. Real-life examples are chemical reactions in CSTRs, cells, plants, and even humans. Respective applications may include the optimization of yield or selectivity in chemical processes depending on available composition of feedstock, optimal control of biotechnological production processes to maximize the yield while maintaining the required quality of therapeutic proteins, the optimization of genetics of plants to maximize biomass production, or the optimization of therapeutical approaches of complex diseases according to the individual disposition of the patients.

Control and optimization of the system can be achieved by trial and error or by the educated guess of scientists and engineers. However, especially if the set of sensitive input variables is large, the dimensionality of the variables to be taken into account leads to tremendous experimental efforts and optimization results that are often far from the true optimum. It is obvious that the knowledge of the system function F could dramatically reduce the experimental efforts and could lead directly to the optimum to be achieved for the applications. Therefore, an efficient modeling platform enabling the development of reliable models describing the output z depending on all relevant input variables x in a reasonable range of values is crucial for successful problem solving.

Establishing an efficient modeling platform in a traditional manner depends on a detailed understanding of the mechanisms controlling the system behavior. In many applications in real life, such as industrial chemical processes using catalysts, cultivation of mammalian cells, or pharmacology, we are far away from a detailed understanding of these mechanisms. Therefore, it is often impossible or not affordable to establish the system function F based on mechanistic understanding.

However, to control or to optimize the output z of the system successfully, the inner mechanisms of the process do not necessarily need to be understood in detail. In these cases, the realization of the model F is not necessarily based on the detailed knowledge of underlying mechanisms of the process, but instead can reasonably be achieved by the extraction of the necessary information from data alone, using any *learning* algorithm. This type of model is referred to as a *data-based model*, in contrast to mechanistic models representing the process mechanisms.

4.3 Principles of Data-Based Modeling

In this chapter, the focus is on the analysis of applications where the system does not explicitly depend on time and the output follows changes of the inputs without delays. All changes in the inputs can be assumed to be slow enough to keep the systems near a steady state (quasi-steady-state assumption).

The model F can thus be defined as follows:

Definition 1: Let $\Omega \subseteq \mathfrak{R}^n$ be the admissible range in the space of the input variables $x \in \mathfrak{R}^n$. This means that the input variables of the process can only be controlled within Ω. Therefore, the behavior of the process outside Ω is not relevant for our application. Then, *a data-based model F* is an algorithm allowing the calculation of the output z of the system for all $x \in \Omega$.

Definition 2: The *realization* of the model F means a mathematical representation of F in terms of a function $F' \in H$, where H is an a priori chosen function space, such that $\|F'(x) - z\| \le \varepsilon \ \forall x \in \Omega$, with ε sufficiently small and $\|\cdot\|$ chosen as the absolute norm.

NOTE: In contrast to mechanistic models, where the functional representation is determined by the underlying, fully explored mechanisms of the system, in data-based modeling the function space H used for the realization of the model F can be chosen according to the preference of the modeler. Therefore, the realization of the model (which is free of choice) and the model itself (which is determined by the (unexplored) system) have to be separated.

To perform data-based modeling in practice requires both: On the one hand, it requires a profound knowledge about the structure of R and the functional patterns that can be covered by the respective approximation subsets with small k. On the other hand, similar knowledge about the system is required that aids in the decision of whether approximations with low complexity—for example, the number of parameters in the realization R—will fit the input-output properties of the system.

For linear realizations R (approximation by sums of orthogonal function systems) the required knowledge is established. In contrast, for nonlinear realizations R (like general neural networks or sparse grids) the characterization of the respective patterns is much less well established.

An important exception for a nonlinear realization for models describing the behavior of systems with many input parameters and nonlinearity are neural networks with feed-forward structure and sigmoidal transfer functions of the nodes. For this realization, Barron's theorem (Barron 1994) uniquely characterizes the functional patterns where the number of parameters increases only polynomially with the number of input variables. Fortunately, lots of applications in process engineering and material sciences appear to satisfy this requirement, although a wide area of applications fails to do so.

4.4 Structured Hybrid Modeling: Introduction

4.4.1 The Data Base as the Generic Concept of Structured Hybrid Modeling

A significant improvement of the drawbacks of data-based modeling can only be expected by a systematic integration of the data-driven modeling approach and additional a priori knowledge about the system behavior. In practice, additional a priori knowledge about the system is mostly available but omitted in the purely data-driven modeling approach.

In the following section, the systematic integration of knowledge in the data-driven modeling process is analyzed. Since the a priori knowledge is often far from complete, purely mechanistic modeling approaches are often not applicable. Therefore, an intermediate modeling approach must be adopted by systematic integration of all available knowledge such that a realization of the model can be accomplished with low number of data samples N_d required for model identification, which will be called structured hybrid models (SHM). The identifiability by using data that are spread around a low-dimensional surface in the high-dimensional data space is essential for SHMs. If that is the case, highly correlated data, typically available in many applications, will be sufficient for model identification.

Suppose a system is described by n independent input variables x and one output variable z, and that the input-output function is assumed to be smooth in x. Moreover, a priori information I about the system is assumed to be available, and it allows the establishment of an abstract *information flow IF* on Ω, which generates the following property of the model F':

Definition: Suppose the set of experimentally available data $D =: \{(\underline{x}_1, z_1), \ldots, (\underline{x}_i, z_i), \ldots, (\underline{x}_{N_D}, z_{N_D})\}$ is sufficiently dense in an open δ—environment of Γ, where $\Gamma \subset \Omega$ is a manifold imbedded in Ω. Then Γ is a data base of the model, if the data on Γ combined with IF allow the reconstruction of F' on entire Ω such that at least the graph $G(F')$ of F' is unique on Ω:

$$\left\{ D\big|_{U_\delta(\Gamma)}, IF \right\} => G\left(F'\right)\big|_\Omega \text{ is unique.}$$

It is not required to identify F' without invariants if the invariants are neutral with respect to the input-output relation of the model.

First the data bases are discussed with the structure of a differentiable manifold and its consequences:

1. For any data-based model, the admissible set Ω itself is a (trivial) data base.

2. If a data base Γ exists with $\dim(\Gamma) = n_\Gamma < n$ and Γ is a differentiable manifold, then the upper bound for the data demand $\bar{N}_d$ can be given by

$$\bar{N}_d \leq C(\|\Gamma\|)\frac{1}{\varepsilon^{n_\Gamma}}$$

3. In the case of mechanistic modeling, where only a fixed, finite set of n_p parameters has to be identified from the data, it holds $\bar{N}_d = n_p$ not depending on the requirements on the interpolation error ε of the model. Therefore, the data base of all mechanistic models consists of n_p isolated points in Ω and has the dimension zero.

    Notice that both in mechanistic and in data-based modeling, the data base is trivial and it does not need to be considered.

4. If a data base Γ exists with $\dim(\Gamma) = m < n$ and Γ is a differentiable manifold, then the model can be partially extrapolated: since Γ is a manifold in Ω with at least co-dimension 1, there is an open environment $U(\Gamma)$ of Γ (larger than $U\delta$ (Γ)), such that F' can be extrapolated from Γ (where the data are measured) onto $U(\Gamma)$ (where no data have been measured to train the model).

    Now model structures need to be found that generate an information flow *IF* allowing the establishment of a low dimensional manifold Γ as a data base.

4.4.2 Functional Networks as an Example of Structured Hybrid Models

As a model class where the data base concept has been studied in detail, the so-called functional networks are a model class where the database concept has been studied in detail. These functional networks, consisting of a network of nested functions—each representing a real subsystem of the overall system—will be discussed here.

Functional networks can be seen as a complement of neural networks: in neural network modeling, the system is modeled by a network connecting the input variables and the output variable by a graph, which is characterized by an unknown structure S, where each node transforms its inputs to the output value of the node using nonlinear transfer functions of the same type (e.g., tanh functions) for all nodes. The identification of the model is carried out by optimizing the structure S of the network together with very few node and edge parameters. This high flexibility in the structure S allows the neural networks to approximate any function depending on n variables (Rojas 1996).

In contrast to the neural networks, functional networks have an a priori given structure S, but the transfer functions may be different in each node and structurally unknown. Therefore, functional networks can be represented by a given network structure S connecting the input variables x to the output variable z. The action of each node of the network is characterized by a transfer function, which is not of the same type throughout the entire network. The choice of the transfer function for each node is free. For example, the transfer function of a node can be a data-driven model itself. In contrast, if a node represents a functional unit of the system for which a mechanistic model is already available, the transfer function of the respective node can be represented by the respective mechanistic model.

Therefore, if no mechanistic model for any node is established, the realization of the functional network model can be a network consisting of lots of neural networks (with hopefully a low number of inputs each) instead of one neural network for the entire process with a large number of input variables. In contrast, if mechanistic models for all nodes are established, a purely mechanistic model results.

Such functional networks can always be established if the system to be modeled consists of subsystems and the connections between the subsystems are known. A range of successful applications (Schuppert 1996; Mogk et al. 2002; Schopfer et al. 2005; Kahrs and Marquardt 2007) have been reported, and a systematic implementation and detailed industrial examples will be presented in Section 4.6.

The analysis of the properties of functional networks goes back to Hilbert's 13th problem, which has been solved by the seminal work of Kolmogoroff, Arnol'd, and Vitushkin (Vitushkin 1954). They established and mathematically proofed criteria allowing the decision of whether functional networks can represent all functions depending on n variables or only a constrained set of functions but did not discuss it in the framework of modeling technology.

Recently it has been shown (Fiedler and Schuppert 2008; Schuppert 1999) that for all functional networks where S has a tree structure, there exist data bases Γ with a minimal dimension equal to the maximum number of input edges of any black-box node in the network. Moreover, it could be shown that almost all differentiable, monotonic submanifolds $M \subset \Omega$ with $\dim(M) =$ maximum number of input variables in a black-box node have (at least locally) the properties of a data base. Additionally, direct and indirect identification procedures have been analyzed and implemented in software (Schopfer et al. 2005).

The generation of an information flow by the functional network structure S is discussed using the example of functional networks with a tree structure.

4.4.2.1 Functional Network with Tree Structure: Basic Examples

Let us start with the simplest example,

$$z(x,y) = u(x) + v(y), x \in \mathfrak{R}^1, y \in \mathfrak{R}^1, z \in \mathfrak{R}^2 \to \mathfrak{R}^1 \tag{4.1}$$

where u and v are unknown functions depending on only one variable each. The functional network, which represents the output z, has the tree structure shown in Figure 4.1a.

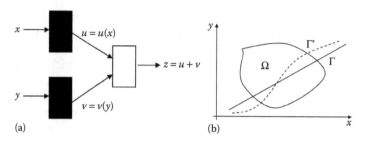

FIGURE 4.1
(a) The structure of the most simple hybrid model with nontrivial data base; (b) two examples of admissible data bases for the SHM structure depicted in Figure 4.1a.

According to the functional structure of Equation 4.1, partial differentiation of z with respect to x and y shows that, whatever the differentiable functions $u(x)$ and $v(y)$ will be, z will satisfy the hyperbolic PDE

$$z_{xy} = 0 \qquad (4.2)$$

where z_{xy} means the mixed partial derivatives of z with respect to x and y. This is obvious, since Equation 4.1 is the general solution of Equation 4.2. This means that the functional network with a structure S given by Equation 4.1 can only describe functions $z(x, y)$ that are the solution of a PDE Equation 4.2, a result that depends only on our a priori knowledge about the functional structure S and does not depend on any data. Obviously this structural knowledge generates an information flow IF: it is fully sufficient to measure the data around a curve $\Gamma =: \Re^1 \rightarrow \Re^2$, $\Gamma \subset \Re^2$ such that the initial value problem for the solution of the PDE Equation 4.2 is well posed. This means, that, if we estimate the set

$$\left\{z, z_x, z_y\right\}\big|_\Gamma$$

from the data, then the solution of Equation 4.2 is unique around Γ. In this example it is obvious that for each Ω there exists an infinite set of curves Γ such that Ω is fully covered by the area of unique solubility of Equation 4.2 given initial values on Γ (Figure 4.1b).

In this example even the measurement of data leading to a uniquely identifiable output z of the system may not lead to a unique identifiability of $u(x)$, or $v(y)$: z is invariant with respect to the transform

$$u \rightarrow u+c, \quad v \rightarrow v-c$$

such that u and v can only be identified up to a (joint) constant c. This invariant, however, has no impact on the graph of $z(x, y)$ we aim to model.

The basic idea in the example Equation 4.1 was that the a priori knowledge of S alone induces an information flow in terms of a partial differential equation for z. Therefore, the functional network can no longer represent any function depending on two variables, but only a small subset.

This idea can be generalized to functional networks S with a tree structure, where all nodes represent subsystems with unknown input-output mechanisms for all nodes. Then all the transfer functions of all nodes have to be represented by a black-box model.

The basic idea can be seen in the example

$$z = w\left(u\left(x_1, x_2\right), v\left(y_1, y_2\right)\right), u = \mathfrak{R}^2 \rightarrow \mathfrak{R}^1,$$

$$v = \mathfrak{R}^2 \rightarrow \mathfrak{R}^1, w = \mathfrak{R}^2 \rightarrow \mathfrak{R}^2, z = \mathfrak{R}^4 \rightarrow \mathfrak{R}^1 \tag{4.3}$$

where all functions u, v, w shall be unknown; we only assume them to be smooth and monotonic.

The respective functional network structure S is depicted in Figure 4.2.

Applying the chain rule, partial differentiation of Equation 4.3 with respect to x_1 and x_2 leads to the following equation system:

$$z_{x_1} = w_u u_{x_1} \tag{4.4}$$

$$z_{x_2} = w_u u_{x_2} \tag{4.5}$$

Note that in both equations the left side can be estimated from the data, whereas the right-hand side depends only on the given structure of S.

Dividing Equation 4.4 by Equation 4.5 allows the elimination of the term w_w which is the only term consisting of elements containing the y-variables:

$$\frac{z_{x_1}}{z_{x_2}} = \frac{w_u u_{x_1}}{w_u u_{x_2}} = \frac{u_{x_1}}{u_{x_2}}\left(x_1, x_2\right) \tag{4.6}$$

Therefore, all derivatives of Equation 4.6 with respect to y_1 or y_2 will vanish, such that each z that can be represented by Equation 4.3 has to satisfy the equation system

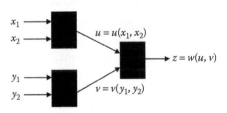

FIGURE 4.2
The simplest structure for a functional network with no known functional nodes and a nontrivial data base.

$$\partial_{y_j}\left(\frac{z_{x_1}}{z_{x_2}}\right)=0,\forall j=1,2$$

Therefore, z has to satisfy the set of PDEs

$$z_{x_2}\partial_{y_1}z_{x_1}-z_{x_1}\partial_{y_1}z_x=0$$

$$z_{x_2}\partial_{y_2}z_{x_1}-z_{x_1}\partial_{y_2}z_{x_2}=0$$

$$z_{y_2}\partial_{x_1}z_{y_1}-z_{y_1}\partial_{x_1}z_{y_2}=0$$

$$z_{y_2}\partial_{x_2}z_{y_1}-z_{y_1}\partial_{x_2}z_{y_2}=0 \tag{4.7}$$

Equation 4.7 show that all possible models F' representing the overall behavior of the system have to be a solution of four hyperbolic PDEs. Therefore, it is sufficient again to measure the data around a manifold such that they provide initial conditions for a unique solution of the equations system Equation 4.7. It has been shown (Schuppert 1999; Fiedler and Schuppert 2008) that Equation 4.7 generates an information flow that is sufficient to guarantee that any two-dimensional, sufficiently differentiable, and monotonic manifold $\Gamma=:\mathfrak{R}^2\rightarrow\mathfrak{R}^4$ is a data base for the system Equation 4.3. It should be noticed here that, in contrast to the first example, Equation 4.1, the extrapolation range for the example Equation 4.3 (which is the area where the existence of a unique solution of Equation 4.7 around a given manifold Γ of initial conditions can be guaranteed) around a manifold Γ can no longer be predicted a priori. The extrapolation range can, however, be calculated *a posteriori*.

4.5 Practical Realization of Hybrid Models

A hybrid model consists of a combination of black-box models (e.g., artificial neural networks) with rigorous models and their connection structure. Therefore, the black-box models are not based on a given structure but contain instead an a priori unknown number of parameters. Usually the structure of the rigorous models is fixed and they may or may not contain parameters. To identify the right number of parameters and their values from given data, specific numerical strategies have to be used.

In this section an overview is provided of efficient methods for parameter identification of hybrid models and of methods for model discrimination to find the right number of parameters for the black-box submodels. Particularly in the case where the hybrid model contains more than one black-box submodel, the latter part could lead to a combinatorial explosion of computation time. In such cases, pragmatic strategies have to be applied.

4.5.1 Parameter Identification with Fixed Number of Parameters

Suppose the hybrid model

$$y = F(x,p)$$

$$p \in \Re^m$$

with black-box submodels of fixed complexity has to be fitted to the given data

$$(x_j, y_j), j = 1,..,L$$

$$x_j \in \Re^n, y_j \in \Re$$

Then the following minimization problem has to be solved:

$$H(p) \rightarrow Min, \ p \in R^m \quad p_{r,min} \le p_r \le p_{r,max}, r = 1,...,m \tag{4.8}$$

$$H(p) = \sum_{j=1}^{L} w |r(F(x_j,p)) - r(y_j)|^q$$

with suitable chosen weights w_j and scaling function $r(x) = x$ or $r(x) = \ln(x)$, where $q \ge 1$. In case $q = 2$, this becomes the least-square version of the minimization task Equation 4.8. A necessary condition for a minimum of Equation 4.8 is

$$G(p) := \frac{\partial H}{\partial p}(p) = 0, \ p, G \in R^m \tag{4.9}$$

Using Newton's method for solving Equation 4.9 would require the calculation of the Hessian of H. Usually this would be very time consuming, and in addition the inverse of the Hessian in many cases (if we have many unknown parameters) would not exist. Hence, similar to the situation of pure black-box models, this is not a suitable method for identifying the unknown parameters of a hybrid model.

4.5.1.1 Levenberg-Marquardt Method

By setting $q = 2$, $w_j = 1$ and $r(x) = x$ and linearization, we get

$$H(p+\delta p) = \sum_{j=1}^{L} (F(x_j,p+\delta p)-y_j)^2 \approx \sum_{j=1}^{L} (F(x_j,p)+J\delta p - y_j)^2$$

$$= \left\| F(x,p)+J\delta p - y \right\|^2$$

with the Jacobian J of F. H is minimal if δp satisfies

$$J^T J\delta p = J^T \left(F(x,p)-y \right)$$

which is the so-called *Gauß-Newton method*. As a damped or regularized version

$$\left(J^T J + \mu I \right)\delta p = JT\left(F(x,p)-y \right)$$

this becomes the so-called Levenberg-Marquardt method (Marquardt 1963) ($\mu > 0$). For small values of damping factor μ, this approaches the Gauß-Newton method, whereas for large values of μ, the resulting step δp is approximately in gradient direction. The Levenberg-Marquardt method is in many situations stable and fast enough to find good solutions of Equation 4.8. In practice (depending on initial values), several hundred iterations are needed to find suitable solutions. Since it is a local convergent method, it has the drawback of possibly stepping into a local minimum.

Other suitable methods to solve the minimization task Equation 4.8 are the derivative-free *Nelder-Mead* or *downhill simplex method*, variants of *gradient descent* (e.g., *back propagation* in the case of purely black-box models), or *evolutionary algorithms*. All three types of algorithms are usually very stable but slow, often requiring several thousand iterations.

4.5.1.2 Remarks

- Usually *early stopping* is used as a criterion to terminate the numerical iteration and to avoid overfitting. By that means a split into a training and a validation set of the available data is used. The training set is then used to minimize Equation 4.8 with a suitable previously mentioned optimization method. The prediction error of the model on the validation set is considered to control the model behavior on unknown data. For example, the iteration procedure could be terminated if the model error on the validation set increases. As an alternative, a maximum number of iterations is performed, and the parameter values corresponding to the smallest validation set error are returned as the best solution.

- To circumvent the problem of local minima, the numerical minimization procedure could be run several times with different initial conditions. Possible choices for initial values are, for example, the use of random values or to perturb the best solution so far (e.g., with a factor of 1%–100%). Again, the behavior on the validation set could be considered to select the optimal parameter vector.

- One reason for local minima is the missing natural order of the inner neurons in the black-box submodels: any permutation of the neurons within one layer results in an equivalent model of the same prediction quality. Symmetry breaking transformations of the parameters of black-box submodels can be used to avoid this and can lead to a huge reduction of the solution space.

- Scaling of input and output parameters is often helpful for improving convergence behavior and balancing error behavior in the case of several model outputs.

4.5.2 Parameter Identification with an Unknown Number of Parameters and Model Discrimination

Hybrid models usually contain at least one black-box submodel that has an a priori unknown number of unknown parameters. Finding the right number of parameters of the black-box submodels is a difficult but crucial task: If the complexity of the black models is too low (i.e., the number of free parameters is too low), the model usually underfits the data. On the other hand, if the model complexity is too high (i.e., too many free parameters), this results in overfitting of the data (Figure 4.3).

A systematic method commonly used in data-based modeling is to split the data into training, validation, and test sets. The usual procedure is as follows:

FIGURE 4.3
Examples of data-based models with underfitting (left), overfitting (middle), and adequate fitting (right) behavior.

- Choose the complexity of the black-box submodels.
- Perform the iteration method to minimize training set error.
- Calculate the validation set error on selected iteration steps.
- Iterate until the maximum number of iterations or another criterion for termination is reached.
- Choose the model with minimum validation set error (Figure 4.4).
- Use the test set to assess the model on an independent data set.

Often, 60% of the data is taken for training and 20% for validation and test sets, respectively. If the data are taken from time series or are specifically distributed in the data space (e.g., clustered), the selection of the validation and test sets has to be carried out with particular care. On the one hand, the validation set data should not be nearly identical to the training set; on the other hand, it should not be completely different (e.g., lying in another part of the data space).

Standard strategies for choosing the validation and test sets are as follows:

- Use every n-th data point (e.g., n = 5; if data are well distributed, no time series data)
- Block-wise (e.g., 20% of the data from a connected time period for validation and test sets, respectively; e.g., for time series data)
- Random (e.g., 20% at random for validation and test sets, respectively)

To avoid dependency on a specific choice of the validation set, it could be varied and the model complexity chosen such that the models perform best on the average of different validation sets (cross-validation).

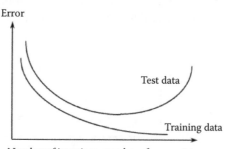

FIGURE 4.4
Typical behavior of prediction error on training and validation, depending on model complexity and number of iterations.

Possible cross-validation strategies are (Hastie et al. 2001) these:

- Use every n-th data point as validation set, starting with first, second ..., n-th data point
- Block-wise validation set; vary the block n times
- Random validation set; choose random validation set N times
- Leave-one-out strategy

4.5.2.1 Model Complexity Variation

The complexity of the hybrid model increases with the complexity of the black-box submodels. In the case where these black boxes are artificial neural networks with one hidden layer (which are universal approximators of continuous functions), the number of hidden-layer neurons determines the complexity of the black boxes. To find the appropriate complexity, the number of hidden neurons is usually varied between fixed upper and lower bounds. The upper bound could be increased with increasing number of data points. Often the variation on this procedure is carried out by linearly or exponentially (e.g., by doubling) increasing the number of neurons.

If the hybrid model contains more than one black-box submodel, identifying the suitable model complexity becomes a combinatorial task. The simplest but most expensive method (in the sense of computing time required) is to try out all combinations of possible complexities of the submodels (i.e., all possible combinations of black boxes with hidden neurons between given minimum and maximum values in the case of artificial neural networks with one hidden layer). Other strategies are to choose the number of parameters randomly or to use an evolutionary algorithm to optimize the number of parameters.

One often very efficient strategy is the following greedy method:

- Start with a model with the lowest possible complexity (all submodels with the lowest number of parameters).
- Fix one submodel where to vary the number of parameters, keep the number of parameters for all other submodels fixed.
- Increase the number of parameters of submodel between minimum and maximum value.
- Choose the version with the best validation set behavior.
- Go to the next submodel.
- Choose the version with the overall best validation set behavior.

This method should be carried out in such a way that the overall complexity increases from a version where the number of parameters of all submodels is at the minimum value to a version where all the parameters are at their maximum value.

4.5.2.2 Model Discrimination

In cases where the available data set is small, other (validation set free) methods could be used to find the appropriate complexity of black-box submodels. These methods usually use a combination of the prediction error $H(p)$ and the number of unknown parameters m to compare models of different complexity and to balance the prediction error and the model complexity (Hastie et al. 2001).

Possible information criterions are as follows (here L is the number of given data points):

- Akaike (based on Kullback-Leiber information) (Akaike 1973): Choose the model complexity according to the model with minimum $AIC = L^*\ln(H(p)/L) + 2m$
- Bayes (Schwarz 1978): Choose the model complexity according to the model with minimum $BIC = L^*\ln(H(p)/L) + \ln(L)^*m$
- For values of $L > 7$, BIC usually prefers models with fewer parameters than for AIC.

4.5.2.3 Topology Optimization Using L_1-Penalization

One major drawback of the previously mentioned methods for topology optimization of hybrid models—especially in the case where the model contains more than one black-box submodel—is the number of models to be checked in order to identify the optimal one.

Continuous selection methods start with a sufficiently complex model (in our case, a hybrid model with complex black-box submodels) and include an extra term in the optimization function that prefers smaller values of the weights. For artificial neural networks with weight vector p, the squared L_2-norm of the parameter vector as weight decay term has found widespread use (Ripley 1996). In the context of linear models, this approach is well known as *ridge regression* (Hastie et al. 2001). The disadvantage of using a squared norm is that this formulation tends to shrink all parameters but sets none of them exactly to zero. Therefore, after optimizing the weight decay extended error function, a threshold is chosen and all weights below the threshold are set to zero. Again in the context of linear models, this problem is dealt with by replacing the squared L_2-norm by the L_1-norm of the parameter vector, an approach known as least absolute shrinkage and selection operator (LASSO). Using the L_1-norm leads to estimated parameters being exactly zero. Therefore, this approach can be used to perform model selection in the linear model case.

In Görlitz et al. (2010), this approach has been transferred to artificial neural networks. The application to hybrid models is straightforward. The main disadvantage of the L_1-penalty

$$\arg\min_{p} H(p) + \lambda \left| p \right|_{L_1}$$

compared to the L_2-penalty

$$\arg\min_{p} H(p) + \lambda |p|^2_{L_2}$$

is that it is much more difficult to solve the underlying minimization problem. In the case of linear models and L_2-penalization, the optimization problem can be reduced to the solution of a linear system, whereas for the L_1-penalization, a nonlinear system has to be solved. Using L_1-penalization for hybrid models leads to an optimization function that is only piecewise differentiable, with a discontinuity at value 0 for any parameter p_j. Hence, usual optimization algorithms run into difficulties as some parameter values approach zero. Therefore, in Görlitz et al. (2010), a modified version of the Levenberg-Marquardt method is used to deal with the non-differentiability of the optimization function in a fashion similar to the approach introduced in Osborne et al. (2000) for the case of linear models. A detailed discussion of related problems with discontinuities in the derivative can be found in Overton (1982).

A problem with using L_1-penalization is the identification of the value of the trade-off parameter λ. One way is to use a validation set to identify the optimal value. Although this approach again results in subsampling of the data, it has the advantage that only one parameter (λ) instead of several complexity parameters for all black boxes has to be optimized.

4.6 Applications

4.6.1 Optimization of a Metal Hydride Process

The simplest hybrid structure consists of a combination of a white-box submodel with a black box to describe the quality properties of a product (Figure 4.5).

The main difference is that instead of using a pure black-box model for the whole relationship, process knowledge could be used to reduce the dimensionality of the input into the black-box model.

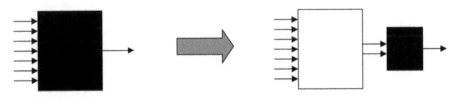

FIGURE 4.5
Black-box model and simplest hybrid model structure.

Important quality parameters for rechargeable batteries are their capacity and the number of possible charging cycles without loss of capacity. Such requirements result in corresponding requirements on the intermediates. In case of spherical metal hybrid intermediates, for example, for NiMeH-batteries, a complex structure optimization on different levels of the spherical agglomerates is necessary: In particular, the size of the primary particles, the crystallites, should be as small as possible. On the other hand, the spherical agglomerates should be as large as possible.

To minimize wastewater that is contaminated with several heavy metals, the main part of the mass flow is reused during an industrial production process of the metal hybrid intermediates (Figure 4.6).

To understand the relationship between process parameters such as temperature, NH_3, NaOH, dopants concentration, and so on, and the resulting product quality (e.g., crystal size, particle size distribution, etc.) for this production process, a hybrid model of the previously mentioned structure is applied. Chemical process knowledge could be used to combine three of the process parameters (temperature, NH_3, and NaOH) with two intermediate parameters A and B. Since it is known that the product quality directly depends only on these intermediate parameters (and potentially on additional process parameters, such as dopants concentration), it is possible to reduce the dimensionality of the black-box input vector by 1 (Figure 4.7).

The left-hand side of Figure 4.8 shows the dependency of the crystallite size and the D_{50}-value of the particle size distribution on NH_3 concentration and temperature for fixed values of all other model input parameters. The model allows the prediction of parameter regions where sufficiently good product quality can be expected. With regard to an energy-efficient process

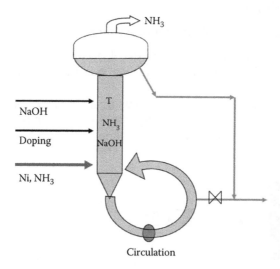

FIGURE 4.6
Simplified scheme of an industrial production process for spherical metal hydride oxide.

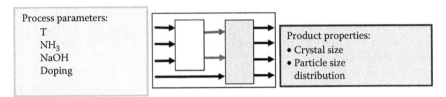

FIGURE 4.7
Hybrid structure of the metal hydride oxide production process.

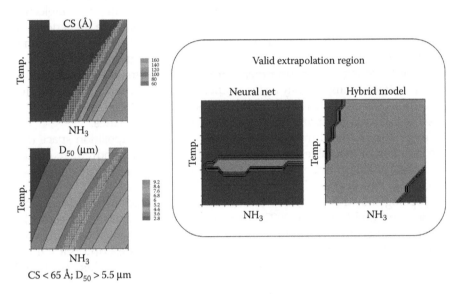

FIGURE 4.8
Dependency of crystal size (CS) and D_{50}-value of the particle size distribution on input parameters (left-hand side). Regions with good product quality are shadowed grey (e.g., CS < 65Å and D50 > 5.5 μm). Valid prediction range (bright grey region) for the black-box and the hybrid models are illustrated on the right-hand side.

operation, the high-temperature regions are of specific interest. Since experimental data were available just for a small (lower) temperature region during the process development phase and pure black-box models cannot extrapolate outside the given range of input data, a neural network model would not have allowed a reliable determination of suitable parameters with respect to product quality for high-temperature regions. For the hybrid model, the suitable extrapolation range (e.g., in the temperature, NH_3, and NaOH space) is given by all combinations of the input parameters such that the intermediate parameters A and B are within the region covered by training data for the black-box submodel. Due to the fact that the intermediate parameters A and B depend only weakly on the temperature, this leads to a tremendous increase of the valid model prediction range of the hybrid model compared to a pure black-box model (right-hand side of Figure 4.8). The process behavior at higher temperature regions was successfully predicted with the help of the hybrid model and was later verified during process operation.

4.6.2 Modeling Complex Mixtures

The principal feature of fragrances and flavors is the high number of ingredients used in their composition. Typical perfume formulae can contain 50 or more ingredients in which each ingredient may be an individual molecule, a mixture of isomers, or a mixture of ingredients. A complete breakdown of these formulae may therefore yield hundreds of chemical species. On the other hand, perfumers often mention the existence of synergies and antagonisms between these ingredients. In other words, the performance of perfumes in terms of olfactory impact may be the result of nonadditive effects. Nonadditive (nonlinear) effects such as flash point, viscosity, or interfacial tension are also observed when considering the physical properties of mixtures.

For perfumes, the dimensionality of the mixture composition space (i.e., the number of available ingredients) is often several hundred, requiring an unreasonable number of experimental data for successful data-based modeling. However, each ingredient can also be characterized by molecular topological, structural, quantum-chemical, and electrostatic descriptors (Todeschini and Consonni 2002). This alone would obviously not solve the dimensionality problem, but the mixtures can be represented by using concentration-weighted averages of these descriptors. If m descriptors are used, modeling is then performed in m-dimensional space only, where m should be as small as possible.

In Mrziglod et al. (2010), a hybrid model was applied, including additional structural information on the ingredients. The heart of this approach is the transformation of the perfume database into a form that can be used in the modeling unit—in other words, the conversion of each formulation into a vector of weighted molecular descriptor sets. This is equivalent to running the black-box submodel in the so-called mixture descriptor space (Figure 4.9).

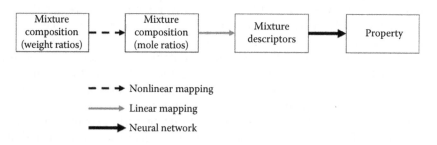

FIGURE 4.9
Hybrid model structure for the modelling of complex mixtures.

In Mrziglod et al. (2010), the composition of the mixtures, values for product quality parameters, and molecular descriptors—which have been calculated with the Molecular Operating Environment (MOE) from the Chemical Computing Group, Inc. (2009)—were fed into the model. The perfume descriptors were calculated by weight-averaging the molecular descriptors for each composition, based on their molar ratios. Then, the number of descriptors was reduced step by step by training an artificial neural network and then eliminating the descriptors that were of less influence on the product quality values. In this way a model was built that needed 10 descriptors per ingredient and mixture.

It was shown in Mrziglod et al. (2010) that nonadditive properties of complex mixtures can effectively be modeled by associating such properties with newly defined vectors containing all the necessary mixture information. These vectors can be defined in the so-called descriptor space with weighted molecular descriptors as coordinates. This procedure has successfully been applied to both qualitative scores and continuous physical properties of perfumes. Furthermore, it was shown in Mrziglod et al. (2010) that reverse engineering, using DoE methods, can be performed in these spaces, leading to the possibility of creating new mixtures with improved properties.

4.6.3 Software Tool Design and Processing Properties

Design and Processing Properties (DPP) is a software tool (Sarabi et al. 2001) that increases planning security for developers of thermoplastic moldings and speeds up time-consuming design steps in the run-up to large-scale production. DPP is based on several hybrid models. The users can define their own requirements profile—for example, thickness, flow path, processing parameters, and machine setups. The model then determines the properties of each molding from this profile.

DPP is divided into several modules. With the design module, the mechanical characteristics of the thermoplastic under consideration can be determined under service conditions. It supplies stress-strain curves and secant modules, gives the loading limits, and plots the influence of temperature. The design module is based on a hybrid model, which consists of a neural network, a principal component analysis (PCA), and a white-box model. Depending on the formulation of the selected material, the part geometry, and processing parameters such as mass and mold temperature, the neural network (together with the PCA transformation) predicts *characteristic points* of the stress-strain curve. The white-box model is an intelligent interpolation method that is able to reproduce the whole stress-strain curve from these characteristic points (Figure 4.10).

With the processing module, it is possible to establish how a specific design can best be injection molded. It calculates key processing parameters such as the filling pressure at various melt temperatures, the metering time as a function of screw speed, and the level of shrinkage during processing. The

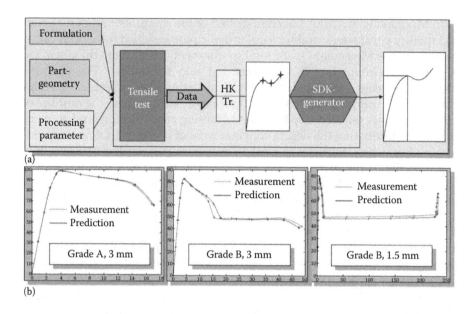

FIGURE 4.10
Structure of the hybrid model for prediction of stress-strain curves (a) and comparison of measured stress-strain curves with curves predicted by the model for typical cases (b).

module is based on a complex hybrid model that contains several black and white-box submodels. The black-box models predict the dependency of the processing properties on the formulation, the part geometry, and the processing parameters that were measured in a very large number of experiments at standard conditions (with respect to machine parameters, etc.). The white-box submodels then transfer these predictions to situations that are more general.

The rheological module provides information on the melt viscosity as a function of shear rate, melt temperature, molecular structure, and fillers. To describe this dependency, the Carreau Ansatz is used. In that way the viscosity η is a known function

$$\eta = f\left(T, \gamma; \eta_0, B, c, TS\right)$$

of temperature T and shear rate γ. The parameters η_0, B, c, TS are modeled as artificial neural networks depending on the formulation of the material.

4.6.4 Surrogate Models for the Optimization of Screw Extruders

Usually, very time-consuming simulations are used for optimizing screw extruders and extrusion processes. In this way, each issue is tackled in a three-stage process. In the first step (pre-processing), the geometry of the screw is determined and converted into a form that allows flow simulations (mesh generation). In the second step (CFD, computational fluid dynamics), the flow simulations are performed by solving differential equations (Figure 4.11). In the third step (post-processing), a comprehensive analysis of the calculated flow fields is carried out in regard to the previous question. However, the procedure is often very inefficient, since a new simulation must be carried out for each question.

A more efficient method for optimizing screw extruders is proposed in Bierdel et al. (2009) by decoupling the simulations of the actual optimization. According to Bierdel et al. (2009), a multitude of simulations in a previously defined parameter space are first performed. The results of the simulations are used to generate a hybrid model for this parameter space. The hybrid model describes the defined parameter space of interest. Based on the hybrid model, predictions for concrete issues can be performed in a much shorter time than would be required for the implementation of a simulation calculation. The data-based model can still be used for the optimization of the extruder parameter values.

In the current situation, three of the parameters predicted by the hybrid model and describing the efficiency of the screw elements are coupled by a known relationship. The resulting hybrid structure is therefore of the form shown in Figure 4.12.

The black-box submodels for A_2 and B_1 are fitted during the model identification such that the data values for A_2, B_1, and B_2 that result from the CFD simulations are met as closely as possible.

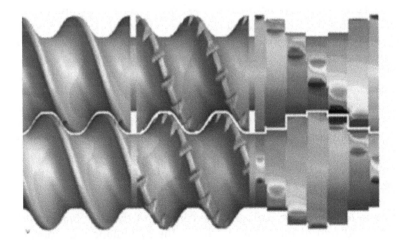

FIGURE 4.11
Visualization of the CFD simulation of the behavior of a screw extruder.

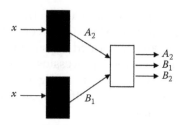

FIGURE 4.12
Hybrid model structure for screw extruders.

4.7 Summary

Due to the widespread availability of numerical tools allowing the proper identification of data-based models from data, the focus of this chapter is not on the learning algorithm, but rather on the requirements of size and structure of the input data, which are necessary to properly identify a model.

For this purpose, the concept of the data base Γ or D (in discrete case) of a model—which is the set of input data sets allowing the identification of the graph of the input-output functions of the system on the entire application space Ω—has been discussed.

The concept of the data base was found to be the appropriate tool to characterize the information content of the a priori knowledge and the benefits of structural knowledge of the system's inner mechanism. It allows the integration of structural knowledge in a systematic and iterative procedure into the data-based modeling process, such that we have today a continuous range in modeling technology connecting the classical *poles*, black box, and mechanistic modeling in a systematic manner. Models in the range between black-box models and mechanistic models in terms of the data base have been called structured hybrid models (SHM). Realization of SHM technology allows the modeler to decide at each stage of the modeling process whether the a priori available knowledge about the system is already appropriate to achieve the modeling goals. Moreover, it provides the tools for a quantitative estimation of the benefits of further knowledge for model accuracy or potentially necessary experimental data acquisition. Therefore, it allows the modeler to establish an optimal, iterative modeling process in which the cost-benefit ratio can be optimized in each step to achieve optimal modeling results from limited resources.

The implementation of the functional structure of the system provides an appropriate approach for realizing the concept of the data base as a basis for structured hybrid modeling. It allows the systematic reduction of the size of the data base of a model, even if no explicit quantitative functional knowledge is available. A possible realization can be achieved by functional networks that provide a straightforward mathematical tool for a systematic crossover from pure black-box modeling to mechanistic modeling with well-defined mathematical properties.

However, there are open mathematical questions remaining. Although the properties of functional networks with tree structures are well established, the properties of more-generic structures are much more complex and are not yet fully discovered. We can provide results indicating that more-complex structures can show properties similar to tree structures, but we are far from a comprehensive theory for generic functional networks. Most probably, an approach focused on application-oriented substructures could be the most appropriate way to realize the performance of the structured hybrid modeling approach for further applications—for example, in systems biology.

Moreover, it is not yet clear whether approaches other than those of functional networks can lead to information flows allowing the realization of dimensional data bases. For example, information flows in terms of elliptic PDEs would allow the reduction of the complexity of a data-based model, similar to the information flows in terms of hyperbolic PDEs as generated by functional networks. However, it is not clear what type of structural knowledge would generate information flows that could be described by elliptic or parabolic PDEs.

References

Akaike, H. 1973. Information theory and an extension of the maximum likelihood principle. In *Second International Symposium on Information Theory*, B. N. Petrov and F. Csaki (Eds.). Budapest, Hungary: Akademiai Kiado.

Barron, A. R. 1994. Approximation and estimation bounds for artificial neural networks. *Machine Learning* 14 (1): 115–133. doi: 10.1007/bf00993164.

Bierdel, M., T. H. Mrziglod, and L. Görlitz. 2009. Data-based models for predicting and optimizing screw extruders and/or extrusion processes. World Patent WO211073181A1, filed December 18, 2009, and issued June 23, 2011.

Fiedler, B. and A. Schuppert. 2008. Local identification of scalar hybrid models with tree structure. *IMA Journal of Applied Mathematics* 73 (3): 449–476. doi: 10.1093/imamat/hxn011.

Görlitz, L., R. Loosen, and T. Mrziglod. 2010. Topology optimization of artificial neural networks using L1-penalization. *20. Workshop Computational Intelligence*. Karlsruhe, Germany: KIT Scientific Publishing.

Hastie, T., R. Tibshirani, and J. Friedman. 2001. *The Elements of Statistical Learning*. Springer Series in Statistics. Montreal, Canada: Springer. Molecular Operating Environment (MOE).

Kahrs, O. and W. Marquardt. 2007. The validity domain of hybrid models and its application in process optimization. *Chemical Engineering and Processing: Process Intensification* 46 (11): 1054–1066. doi: 10.1016/j.cep.2007.02.031.

Marquardt, D. W. 1963. An algorithm for least-squares estimation of nonlinear parameters. *Journal of the Society for Industrial and Applied Mathematics* 11 (2): 431–441. doi: 10.1137/0111030.

Mogk, G., T. H. Mrziglod, and A. Schuppert. 2002. Application of hybrid models in chemical industry. In *Computer Aided Chemical Engineering*, J. Grievink and J. van Schijndel (Eds.), pp. 931–936. Hague, the Netherlands: Elsevier.

Mrziglod, T., L. Goerlitz, C. Quellet, and A. P. Borosy. 2010. Modeling complex mixtures. Research Disclosure No. 556008. Hampshire, GB: Kenneth Mason Publications.

Osborne, M. R., B. Presnell, and B. A. Turlach. 2000. On the LASSO and its dual. *Journal of Computational and Graphical Statistics* 9 (2): 319–337. doi: 10.1080/10618600.2000.10474883.

Overton, M. 1982. Algorithms for nonlinear L1 and L∞ fitting. In *Nonlinear Optimization*, M. J. D. Powell (Ed.). London. Academic Press.

Ripley, B. D. 1996. *Pattern Recognition and Neural Networks*. Cambridge, UK: Cambridge University Press.

Rojas, R. 1996. *Theorie Der Neuronalen Netze*. Heidelberg, Germany: Springer Verlag.

Sarabi, B., T. H. Mrziglod, K. Salewski, R. Loosen, and M. Wanders. 2001. Hybrid model and method for determining the mechanical properties and processing properties of an injection moulded article. Edited by DE50207685D1 DE10119853A1, EP1253491A2, EP1253491B1, US6839608, US20030050728.

Schopfer, G., O. Kahrs, W. Marquardt, M. Warncke, T. Mrziglod, and A. Schuppert. 2005. Semi-empirical process modelling with integration of commercial modelling tools. In *Computer Aided Chemical Engineering*, L. Puigjaner and A. Espuña (Eds.), pp. 595–600. Barcelona: Elsevier.

Schuppert, A. 1996. New approaches to data-oriented reaction modelling. *3rd Workshop on Modelling of Chemical Reaction Systems*. Heidelberg, Germany: IWR.

Schuppert, A. 1999. Extrapolability of structured hybrid models: A key to the optimization of complex processes. *International Conference on Differential Equations (Equadiff)*. Berlin, Germany.

Schwarz, G. 1978. Estimating the dimension of a model. 461–464. doi: 10.1214/aos/1176344136.

Todeschini, R. and V. Consonni (Eds.). 2002. *Handbook of Molecular Descriptors*. H. K. R. Mannhold and H. Timmerman (Series Eds.). New York: Wiley-VCH.

Vitushkin, A. G. 1954. On Hilbert's thirteenth problem. *Doklady Akademii Nauk. SSSR* 95: 701–704.

5

Hybrid Modeling of Biochemical Processes

Vytautas Galvanauskas, Rimvydas Simutis, and Andreas Lübbert

CONTENTS

5.1 Introduction

In recent decades, the extent of research and development involving hybrid systems has increased rapidly. Hybrid systems combining classical first-principles model-based methods, genetic/evolutionary programming algorithms, fuzzy logic, neural networks, and expert systems are proving their effectiveness in a wide variety of real-world problems. Every intelligent technique has particular computational properties (e.g., ability to learn, explanation of decisions) that make them suited for particular problems and not for the others. From the knowledge of their strengths and weaknesses, one can construct hybrid systems to mitigate the limitations and take advantage of the opportunities to produce systems that are more powerful than those that could be built with individual methodologies.

The general aim of bioprocess engineering is to develop and optimize biochemical production processes. Process design, optimization, and control require extensive knowledge of the process. The classical way of representing process knowledge in science and engineering is to use fundamental mathematical models (i.e., based on first principles). These models require a thorough understanding of mechanisms governing the process dynamics. In biochemical processes, however, many essential phenomena are not yet understood in sufficient detail necessary to develop physically based models. Hence, to establish models applicable in practice, additional resources must be exploited (von Stosch et al. 2014a, 2016). Everyday experience shows that a significant amount of quantitative knowledge about the biochemical processes is available that cannot yet be effectively represented in a form of first-principle mathematical models. Thus, it is necessary to look for possibilities to incorporate this knowledge into alternative kinds of numerically evaluable process models by transforming qualitative knowledge to quantitative knowledge. Also, the data from already-running biochemical processes covers a wealth of hidden information about the industrial process. Engineers have been recognizing this and condensing the information within the data in the form of so-called *engineering correlations*. Since this is extremely time-consuming, most data records have not been sufficiently exploited so far. Experience has shown that neither mathematical process models nor heuristic descriptions alone are sufficient to describe real production processes accurately enough so that an efficient automatic control system for a production-scale bioreactor can be based only on this description. In order to overcome this shortcoming, all available knowledge should be utilized. In particular, the information hidden in the extended measurement data records from the process under consideration must be exploited. Hence, procedures are needed to simultaneously capitalize on the available mathematical modeling knowledge, the information hidden in process data records, and the qualitative knowledge gained by process engineers through their experience. This can be achieved by means of the hybrid modeling technique. The hybrid modeling approach is intended for the simultaneous use of different sources of knowledge. It is expected that the utilization of more-relevant knowledge generally leads to improved accuracy of the process model.

This chapter reviews the application of hybrid modeling techniques for biochemical process optimization, monitoring, and control. The methods address important aspects of bioprocess monitoring and control: methods for process/sensor fault analysis, state estimation, open-loop optimization, and closed-loop control.

5.2 Hybrid Modeling for Process Optimization

A straightforward way to improve the performance of industrial biochemical processes is to apply model-based optimization. Such a procedure may lead to the maximization of the target product with respect to the imposed technological and physiological constraints of the culture, assurance of necessary product properties, and minimization of production costs. As the dynamics of biochemical processes is of nonlinear and time-varying nature, and the biochemical phenomena involved are often difficult to explain, their description by means of model-based approaches involving only classical and fundamental models may be difficult to develop. In such cases, application of hybrid modeling techniques may become an alternative to classical modeling approaches (Galvanauskas et al. 2004).

Hybrid models for model-based bioprocess optimization may include basic submodels containing fundamental mass and energy balance equations of a bioreactor-scale system, kinetic rate expressions for reacting components, engineering correlations, artificial neural networks (ANNs), fuzzy expert systems, and other types of models. In biochemical processes, data-driven submodels are successfully applied for modeling of specific reaction rates that include complex kinetics and mass transfer processes (viscosity, mass transfer from gas to liquid phase, etc.). Such data-driven submodels may be used in parallel with the classic submodels by applying weighting functions (*safe models*) (Simutis et al. 1995, 1997).

Application of hybrid models is especially promising when extensive process data is already available from which additional information about the process can be extracted. Unfortunately, the amount of process data may be rather limited if one starts with investigation of a bioprocess that is under development. If the quantity of process data is limited at the beginning of the bioprocess optimization procedure, one starts with relatively simple process models and iteratively increases the model complexity as the knowledge and information about the process becomes more comprehensive due to the data obtained from new runs and various resources. These initial process representations are regarded as a first approach to a well-performing, more complex model since they are based on basic resources of process knowledge—for example, models described in the scientific literature and the knowledge of experienced process engineers working with similar bioprocesses in practice (Lübbert and Simutis 1998). Such an iterative optimization procedure can reduce the number of explorative runs necessary to achieve desired process performance. Depending on the bioprocess model and data available, various structures of hybrid models may be developed (von Stosch et al. 2014b). The main aim of hybrid modeling is to combine the different components effectively, making use of the knowledge fusion. During the procedure, both model parameter identification and model structure adaptation may be applied. A modular modeling concept, where the submodels are arranged in a network structure, might help significantly to increase the performance of the model and to keep its transparency (Lübbert and Simutis 1998). The advantages of such a network structure are manyfold. A transparent structure helps avoid mistakes and simplifies adaptation of the model to the changes in production processes. Different submodels for the same part of the process can be taken from different resources by simply making use of complementary experience and process knowledge. They can be considered as different votes for effective process descriptions, which can simultaneously be used to form a weighted average.

More-precise hybrid models allow for thorough exploring of the state space of the bioprocesses to better predict behavior and make use of these models for more-precise model-based optimization. Application of a hybrid modeling approach for biochemical process optimization is shown in the following sections with regard to several representative examples.

5.2.1 Industrial Penicillin Production Process

The industrial penicillin production process is known for its complexity, especially regarding mass transfer processes and complex kinetics (Lübbert and Simutis 1998). Hence, the classical models of penicillin production processes cannot be used successfully for process optimization, as they are not capable of correctly describing the previously mentioned phenomena. Besides the classical oxygen uptake rate (OUR) model for reaction rates and the mass balance equations, the proposed hybrid model includes two ANN-based models for viscosity and kinetics, respectively, and a radial basis function network (RBFN) compensator for OUR and operator rules (Figure 5.1a). The hybrid model is tuned using a sensitivity function approach and back-propagation technique. After the identification of the hybrid model, a model-based process optimization is performed. The optimization results (Figure 5.1b) show an increased process performance when compared to the results obtained with classical modeling techniques.

5.2.2 Microbial Surfactant Production Process

Another example of hybrid modeling application comes from microbial surfactants production, interest in which has been steadily increasing in recent years due to their wide potential applications in health care, food-processing industries, and environmental protection (decontamination of soil polluted by oil, increasing rate of oil biodegradation by microorganisms, etc.) (Levišauskas et al. 2004). The investigated biosurfactant production process is related to the development of complex technology for cleanup of soil contaminated by oil pollutants. For the implementation of this technology, the production of biosurfactant is required on a large scale,

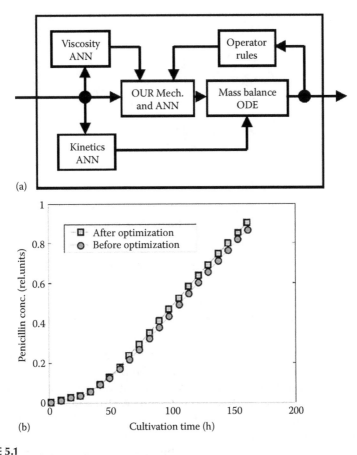

(a)

(b)

FIGURE 5.1
Hybrid approach for optimization of industrial penicillin process: (a) structure of the hybrid model and (b) hybrid model-based optimization results. (Reprinted with permission from Galvanauskas, V. et al., *Bioprocess Biosyst Eng*, 26 (6), 393–400, 2004.)

and therefore the productivity of the production process is, increasingly, a considerable technological problem. The aim of the research was to optimize the process of biosurfactant synthesis in fed-batch culture *Azotobacter vinelandii* 21 in order to maximize the yield of the biosurfactant. For this purpose, a hybrid model was developed and a parametric optimization approach was applied for the optimal feed-rate profile calculation using an RBF network.

Mass balance equations for the principal components and mechanistic equations of specific reaction rates are presented elsewhere (Levišauskas et al. 2004). The main challenge of the process modeling is the development of an accurate model of the specific reaction rates for ammonia and phosphate consumption and for biosurfactant production. Initial investigations

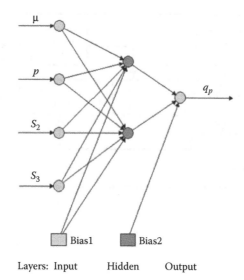

FIGURE 5.2
Structure of ANN for estimation of the specific rate of biosurfactant production. (Reprinted with permission from Levišauskas, D. et al., *Biotechnol Lett*, 26 (14), 1141–1146, 2004.)

have proven that the mechanistic approach does not lead to an acceptable modeling quality due to the complexity of the process. The modeling quality has been significantly improved with the introduction of the black-box submodels for the three previously mentioned specific reaction rates. As an example, the structure of the feed-forward ANN used for product-specific synthesis rate modeling is presented in Figure 5.2. The biosurfactant specific production rate q_p is related to the biomass specific growth rate (μ) and the concentrations of biosurfactant and substrate components. Due to the complexity of the functional relationship qp (μ, p, s_2, s_3), it is expressed by a feed-forward ANN containing 4 inputs, 2 nodes in the hidden layer, and 1 output. The specific consumption rates of ammonia (q_{s2}) and phosphate (q_{s3}) were also modeled by means of the ANNs, containing 2 inputs (μ and q_p), 2 nodes in the hidden layer, and 1 output. The predictions of the three ANNs were incorporated into the respective equations of the mass balance equation system.

Parameter values of the hybrid model were identified using an evolutionary programming algorithm and experimental data from batch and fed-batch cultivation processes. The identified model was validated by predicting unknown experimental data.

The objective of the process optimization was to determine the feed-rate time profile and the nutrient component concentrations in feeding solution that maximize the yield of biosurfactant at the end of the fed-batch process with respect to technological constraints imposed on the feed rate and final cultivation volume.

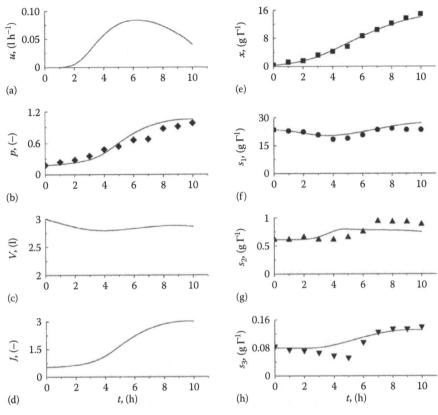

FIGURE 5.3
The optimized time trajectories of the feed-rate and process state variables (solid lines) and measured values of the state variables in the validation experiment (symbols): (a) feed rate; (b) emulsification activity; (c) culture volume; (d) process performance index (total biosurfactant amount); and (e)–(h) biomass, glucose, ammonia nitrogen, and phosphate phosphorus concentrations, respectively. (Reprinted with permission from Levišauskas, D. et al., *Biotechnol Lett*, 26 (14), 1141–1146, 2004.)

Due to the complexity of the mathematical model on which the optimization is based, the feed rate and the feeding solution optimization procedure were implemented using the parametric optimization approach.

The calculated optimal feed-rate time profile and time trajectories of the process state variables are presented in Figure 5.3 (solid lines) along with the experimental data of the optimized process of biosurfactant production in a validation run (Figure 5.3, symbols).

The experimental results demonstrated that the increase of biosurfactant yield due to hybrid model-based optimization totals over 11% as compared to the best result in the initial experiments that were used for the model identification and validation.

5.2.3 Recombinant Protein Production Process

Recently, production of recombinant proteins became an important field of industrial biotechnology. The processes of recombinant protein expression are complex, and the relationships between the key process variables influencing the synthesis have not been very well studied so far. In many cases, an important problem during the optimization of the process is related to the efficient maximization of the soluble protein amount, at the same time minimizing the amount of inclusion bodies (IB) formed during the synthesis process. The development of conventional models based on complex mechanistic expressions including the influence of various process parameters is time-consuming, and the developed models are applicable to the particular process investigated.

A representative example of a recombinant protein synthesis (referred to as HumaX and not specified for confidentiality reasons) in a fed-batch *E. coli* cultivation process and its hybrid model-based optimization is presented in Gnoth et al. (2010).

Before starting with the model-based optimization of the process, it is important to investigate the dependency of the specific product (both soluble proteins and inclusion bodies) synthesis rates on the main control and state variables of the process. In the process under investigation, the main factors influencing the previously mentioned specific reaction rates are the specific biomass growth rate and the cultivation broth temperature.

The investigations have proven that the inclusion body synthesis rate increases with rising temperature, and the synthesis rate of the soluble fraction reaches its maximum at the intermediate temperature of T = 31°C. At a given temperature, the inclusion body synthesis rate increases with the specific growth rate, while the product synthesis rate for the soluble fraction again reaches a clear maximum at the specific growth-rate value of about 0.12 h^{-1} (Gnoth et al. 2010). These observations provided valuable information about the main factors influencing the process performance and the applicable control strategy for the investigated process of the soluble protein production, and, as a result, a more general hybrid model was proposed that was easier to develop (lower model development time costs) and lead to higher modeling accuracy.

The hybrid model (Figure 5.4) is composed of feed-forward artificial neural networks describing the specific rate expressions and mass balance equations. Two separate ANNs were used for the estimation of the specific growth rate (μ) and the specific product synthesis rate (π).

The hybrid model structure is shown in Figure 5.4.

The network for biomass-specific growth-rate calculation has a structure of 5 inputs, 1 bias, 10 hidden-layer nodes, and 1 output, and its inputs are supplied with the substrate feeding rate and total carbon dioxide production rate (*tCPR*) and total oxygen uptake rate (*tOUR*) signals that are available on-line. The estimated biomass-specific growth-rate value at

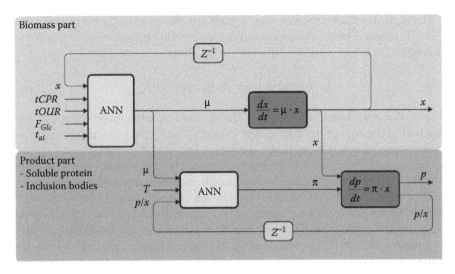

FIGURE 5.4
General structure of the hybrid model describing the specific biomass and product production rates μ and π along with the biomass x and the respective product mass p. (Reprinted with permission from Gnoth, S. et al., *Appl Microbiol Biotechnol*, 87 (6), 2047–2058, 2010.)

the ANN output is passed to the mass balance equation for biomass. At the same time, this signal of the hybrid model, along with the cultivation temperature and the specific protein load p/x, is used as an input in the second ANN for target protein modeling. The structure of the latter contained 3 inputs, 1 bias, 7 hidden-layer nodes, and 1 output. ANNs of the same architecture are used for the calculation of both soluble and inclusion body protein synthesis. For the sake of simplicity, only one ANN for protein modeling is shown in Figure 5.4.

The mass balance equations were solved against the biomass and protein amounts, and not the respective concentrations. The model structure and parameters were identified by means of the sensitivity equation approach (Schubert et al. 1994) and the leave-one-out cross-validation procedure (Hjorth 1994).

The deviations between the analytical measurements of the main process variables (biomass, soluble proteins, and inclusion body proteins) and the computer simulation results showed that the modeling quality using the developed hybrid model is significantly better than the one achieved with mechanistic process models.

After the specific product synthesis rate patterns $\pi(\mu,T)$ have been determined, the trained hybrid model was applied to calculate optimal μ and T time profiles with respect to a defined objective function and constraints.

The optimization problem was solved for the post-induction phase of a fixed duration. The following quantitative objective function was formulated (Equation 5.1):

$$J = p_{Sol}(t_f) - \beta_1 \cdot p_{IB}(t_f) - \beta_2 \cdot x(t_f) \qquad (5.1)$$

where the first term p_{Sol} denotes the desired soluble product mass, [g], reached at the end of the cultivation process ($t = t_f$); the second part p_{IB} takes into account the inclusion body synthesis, [g]; and the last term x, [g], was added to avoid excessive biomass synthesis. The dimensionless parameters β_1 and β_2 can be considered as factors penalizing the value of the performance index J for the synthesis of inclusion bodies and excessive biomass production that does not contribute to target protein synthesis. In practice, excessive biomass amounts would require higher capacities of the bioreactor system equipment, and the higher fraction of IB proteins formed would increase the downstream costs.

When solving an optimization problem, one has to define and meet the constraints that will bind the optimal solution to the feasible limits of the optimized process and equipment. In a real cultivation process, the optimal solution is constrained by technological and physiological factors.

In the investigated process, the main technological constraint is the limited oxygen transfer rate that can be provided by the bioreactor aeration system. In a typical cultivation process, the oxygen demand steadily increases and reaches its maximum at the induction time moment. Therefore, the maximum oxygen transfer rate of a given bioreactor system is closely related to the optimum induction time moment.

From the physiological point of view, the limiting constraint is often the critical specific substrate uptake rate σ_c of the cells, an excess of which leads to the formation of overflow products. In the case of the investigated *E. coli* cultivation process, the main overflow metabolite is acetate. Excessive acetate concentrations inhibit growth, product synthesis, and thus the overall process productivity (Gnoth et al. 2010). As the value of the critical substrate uptake rate considerably decreases with time and there are no reliable models of this phenomenon reported so far, the critical rate was determined experimentally using a modified substrate pulse-response technique based on the approach proposed in Åkesson et al. (1999).

In order to facilitate the implementation of the optimal solution in practical application, some general simplifications were made. After the induction, the temperature was kept constant after ramping down to the optimal value and constrained within the interval $27°C \leq T \leq 35°C$. The second simplification is related to the time profile shape of the specific biomass growth rate. In literature, it was most often assumed that it is necessary to keep the specific growth rate as far as possible constant at its optimal value (e.g., Jenzsch et al. 2006). Some authors report a close-to-linear specific growth-rate profile for optimal soluble protein synthesis that was obtained using analytical optimization techniques and successfully validated experimentally (Levišauskas et al. 2003). In the investigation discussed here, a slight linear shift in μ was introduced, and a linear time function for the specific growth-rate profile was used with bias and slope parameters—a_0 and a_1, respectively.

In order to calculate the optimal μ(t) profile, the parameters a_0 and a_1, and the fermentation temperatures T were varied in order to obtain a minimum in J (Equation 5.1). The results of the numerical optimization depend on the choice of the parameters β_1 and β_2 in the objective function.

During the optimization, various combinations of these parameters were investigated. The parameters were chosen to be $\beta_1 = 1.0$ and $\beta_2 = 0.025$. In that case, the dissolved protein mass was maximal, while the inclusion body formation was kept at a relatively low level (Figure 5.5).

The optimal values of the parameters were $T = 31°C$, $a_0 = 0.12$, and $a_1 = -0.004$.

Figure 5.6 compares the results of the optimized process (run S509) with the ones obtained when a standard control strategy for recombinant proteins was applied (run S400). The latter strategy was proposed at the beginning of the investigation reported (Gnoth et al. 2010), when the strain was delivered

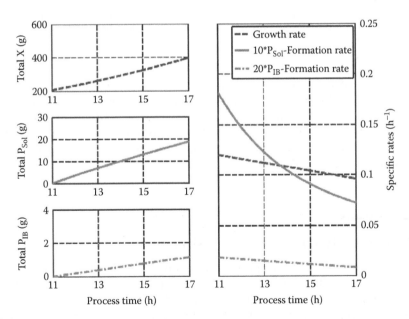

FIGURE 5.5
Simulated time profiles of the process state variables (concentrations and specific reaction rates) of the two optimization scenarios. (Reprinted with permission from Gnoth, S. et al., *Appl Microbiol Biotechnol*, 87 (6), 2047–2058, 2010.)

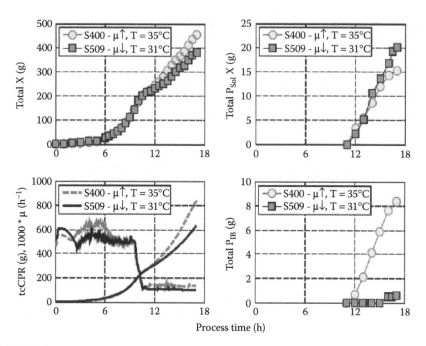

FIGURE 5.6
Comparison of the trajectories obtained with the standard and optimized control strategies. (Reprinted with permission from Gnoth, S. et al., *Appl Microbiol Biotechnol*, 87 (6), 2047–2058, 2010.)

to the lab. It becomes clear from Figure 5.6 that the improvements in process performance obtained as a result of hybrid model-based optimization are considerable. In the optimized case, the results clearly outperform the initial control strategy, as the soluble proteins were produced with lower biomass, the soluble product fraction increased, and the undesirable inclusion body fraction was significantly reduced.

5.3 Hybrid Modeling for State Estimation

In most cases, bioprocess performance and the objective function for process optimization are indirectly determined by the specific biomass growth rate or specific product production rate of the biochemical processes. Hence, the essential practical problem is to determine the relationships for specific rate expressions containing state variables and the variables characterizing the abiotic phase of the culture. Since the structure of the relationships is most often unknown at the beginning of the process modeling phase, artificial neural networks can be used to represent these specific rate expressions. The specific rate expressions can be incorporated in a general bioprocess model represented by mass balance equations, and subsequently the ANN-based specific rate expressions can be identified using experimental data from the actual bioprocess. For the first approach to establish such relationships, one can use all the process variables from which measurement data are available and provide these variables as inputs to the artificial neural network describing the specific rate expressions. However, in order to create more compact relationships for specific growth/production rates and to find the process quantities, which are most important with respect to its influence on the conversion rates, it is better to start the model identification procedure with an ANN, which contains only those variables as inputs. The latter should be a priori known to be of undoubted importance. Then, for one after the other process variable, it can be tested to determine whether its inclusion into the set of ANN input variables leads to a reduction of process modeling error. If the process modeling error can be significantly reduced by the addition of one of the process variables tested, then this variable can be classified as the first candidate to extend the basic model for the specific rate expression. When the reduction in the modeling error is larger than a critical limit of, for example, 5%, the basic model describing the process kinetics can be extended by adding the variable that performed best to its input nodes (Simutis and Lübbert 1997). The structure of such exploration procedure is shown in Figure 5.7.

The classification procedure described previously can be repeated, since cooperative effects between the different process variables could have changed the order of the sequence.

By repeatedly applying this additional classification procedure, all the process variables that have significant influence on the process performance

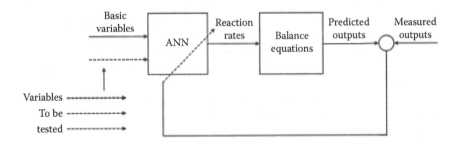

FIGURE 5.7
Scheme of a data exploration procedure to identify the ANN-based relationships for specific growth/prediction rates in cultivation processes. (Reprinted with permission from Simutis, R. and A. Lübbert, *Biotechnol Prog* 13, 479–487, 1997.)

can be incorporated into the process model. This exploration procedure not only provides information about the variables, which influence the key variables of the process, but also creates a data-driven gray-box model that can be used for process state estimation. Moreover, the information obtained about the influence of the different variables considered can be used as a first step in establishing a comprehensive mechanistic process model.

5.3.1 Hybrid Versions of Filters

For more than 50 years, filter algorithms were discussed and used for state estimation, starting with Kalman's seminal paper (Kalman 1960). Filters combine models for the dynamics of the process state (process model) with measurement equations (measurement model). These measurement models relate the state variables, which are often not measurable on-line, with measured variables. In bioprocess engineering, the most important state variables are biomass and product concentrations, and the most important measured variables are the oxygen uptake and the carbon dioxide production rate and the base consumption during pH control (Stephanopoulos and San 1984; Mandenius 2004; Dochain 2008). While the dynamic equations for the state variables are derived from well-established mass balances, the measurement model equations are often based on simple relationships, for example, on Luedeking-Piret-type expressions (Luedeking and Piret 1959). The latter can be replaced by general static functions that can map the biomass to the measurement variables in a much more flexible way, leading to a much higher mapping accuracy. An important example for such a static functional relationship is the artificial neural network. However, the measurement model equations can also be constructed by support vector machines (SVM) and relevance vector machines (RVM) (Simutis and Lübbert 2015).

Since the original Kalman filter can only use linear dynamic and measurement models, extensions of this algorithm must be used. The unscented Kalman filter (Wan and van der Merwe 2001) is a variant that can immediately use complex nonlinear dynamic state equations together with the advanced static measurement model equations in the form of ANNs, SVMs, and RVMs. Another variant of the filters is the particle filter, which also does not require any restrictions to the models. In this way one can easily construct hybrid versions of filters that combine mechanistic dynamic models with modern black-box representations developed by the machine learning community.

A typical example is an estimate, using an unscented Kalman filter, of the biomass and the product concentration in an *E. coli* cultivation producing a recombinant protein on measurements of the cumulative oxygen uptake rate and the cumulative carbon dioxide production rate. The missing restrictions with respect to the measurement model equations allow black-box model

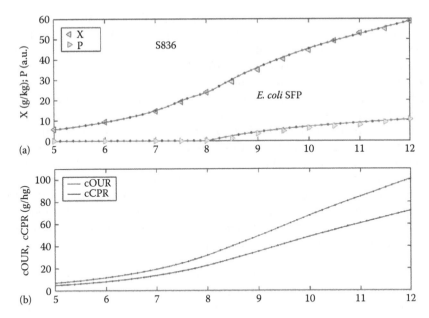

FIGURE 5.8
Estimation of the biomass X and product P concentration with an unscented Kalman filter (a) using the measurements of the cumulative oxygen uptake rate cOUR and the cumulative carbon dioxide production rate cCPR; and (b) an artificial neural network (as a measurement equation) trained on data sets from three fermentations. The symbols are the off-line data that were available only after the cultivations were finished. The lower part shows the measurement data (points) used and their modeled values (lines).

representations of nearly undistorted measurements, particularly the cumulative versions of OUR and CPR. The ANN as measurement equation was identified using data sets from three cultivation runs.

The example (Figure 5.8) shows that the hybrid approach to the state estimation provides excellent accuracies, because practically noiseless cumulative data could be used as measurement signals.

5.3.2 A Fuzzy-Supported Extended Kalman Filter

State estimation requires a sufficiently accurate biochemical process model. Usually models consisting of sets of differential equations are used, and these must be identified from the experimental data of the particular process. The identification of the model parameters is restricted in accuracy by the noise present in the on-line/off-line data obtained from measurements at production-scale cultivation process.

In order to improve the state estimation procedure, the following hybrid method was proposed (Simutis et al. 1992): (a) It is easier to describe a particular process phase than the entire process with a single mathematical model. The model equations can then usually be kept simpler, and the parameters can be estimated with higher accuracy. (b) In order to describe the whole process, a couple of different models for all phases or situations must be used. Where one phase transitions into the next, additional models can be used to describe the process. (c) The advantage in such a piecewise description, however, is offset by the additional efforts to keep the assignment of the different models under control.

The elements by which the idea can be realized are as follows: (a) Already-available data sets of the process, the state of which is to be estimated on-line, can be used in an off-line analysis in order to identify individual process phases that can be described by individual models. (b) Different models can be used in an extended Kalman Filter algorithm for a concrete on-line state estimation. (c) The decision on which model should be used in a particular situation can be based on heuristic knowledge about the process. It can be implemented by means of a compact knowledge-based system. (d) In order to obtain smooth transitions from one partial model to the next one, a fuzzy reasoning technique can be employed with considerable advantage.

Simutis et al. (1992) described how to use a fuzzy expert system in order to select one of a subset of models that may be appropriate to describe the actual process situation. With all these models, the current process state can be estimated using an Extended Kalman Filter (EKF). All individual results can be combined to obtain an effective state estimation using fuzzy reasoning techniques. The structure of the hybrid algorithm used for fuzzy-supported state estimation is shown in Figure 5.9.

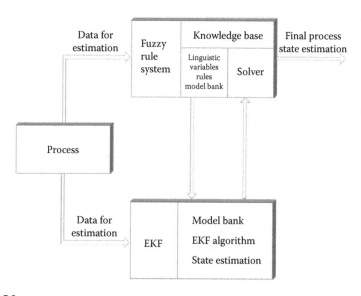

FIGURE 5.9
Structure of hybrid algorithm used for fuzzy-supported state estimation of a biochemical process. (Reprinted with permission from Simutis, R. et al., *J Biotechnol* 24 (3), 211–234, 1992.)

There are two essential advantages of this approach:

1. The first is that the modules can be more closely related to the process dynamics in the corresponding phases and are thus more accurate models.

2. The second is that the models for the individual process phases can be kept smaller and more transparent than in the comprehensive model. The immediate advantage is that it becomes significantly easier to identify the parameters of these partial models. Such a fuzzy-supported state estimator was tested in the industrial beer fermentation process (Simutis et al. 1992). The tests showed that the state estimation using this approach is more accurate than using the usual EKF technique.

5.3.3 Soft-Sensing Methods for Mammalian Cell Cultivation Processes

The U.S. Food and Drug Administration's process analytical technology (PAT) initiative (FDA 2004) has led to a major industry-wide effort to utilize multivariate data acquisition and analysis tools, modern process analyzers or process analytical chemistry tools, process and endpoint monitoring and control tools, and continuous improvement and knowledge management tools (Gunther et al. 2008).

In a biochemical process, one of the key variables to be monitored during the product formation process is the concentration of the viable cells. Several direct measurement devices have been proposed and used to measure the cell number density (Shah et al. 2006). There are several disadvantages of using on-line measurements of viable cell densities in production-scale bioreactors. Most techniques cannot distinguish between viable growing cells and those that are no longer active. Also, they usually require sophisticated equipment that is not easy to validate and maintain in a Good Manufacturing Practice (GMP) environment. Generally, in production bioreactors for mammalian cell cultures, the number of measurement devices is usually kept as low as possible in order to avoid contamination risks and to reduce efforts of sensor validation (Aehle et al. 2010). Indirect estimation using on-line measurements that are already available does not suffer from these disadvantages (Teixeira et al. 2009). However, indirect biomass estimators require models that correlate with other measurements of the viable cell count.

Alternatively, various on-line measurement-based techniques may be used, such as simple Luedeking/Piret and linear approaches using a single measurement signal, multiple linear and nonlinear regressions, principal component analysis (PCA), support vector regression (SVR) and relevance vector regression (RVR) techniques, and feed-forward ANNs.

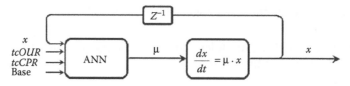

FIGURE 5.10
Hybrid estimator consisting of a mass balance for biomass x and ANN for specific growth μ. (Reprinted with permission from Aehle, M. et al., *Cytotechnology* 62 (5), 413–422, 2010.)

Further improvements in modeling quality can be expected when a pure black-box approach implemented in an ANN is complemented with fundamental knowledge in the form of a mass balance equation (Simutis et al. 1997). Such a hybrid approach is depicted in Figure 5.10. While the biomass balance is defined by the mass conservation law, the kinetic expression for a specific growth rate is less well known. Hence, when sufficient measurement data is available, the kinetics can be represented in a data-driven way. Here the kinetics for biomass estimation is modeled by means of an ANN.

As the direct measurement data for the specific biomass growth rate— which could be used to fit the ANN weights—are not available, the sensitivity approach (Schubert et al. 1994) can be employed to train the network. For this purpose, the measurement data for the input variables of the ANN and the initial value of the cell count for the mass balance equation must be provided.

Generally, all the estimation techniques investigated deliver acceptable results when the measurement data are undisturbed (Aehle et al. 2010). However, a different estimation precision is achieved. The best results with respect to the validation sets were obtained with the hybrid model approach (see Figure 5.11). While the technique requires more-complicated identification procedures, it provides estimates on biomass and its current specific growth rate.

In the same way, it is possible to additionally develop a hybrid estimator for the specific product synthesis rate π. Having both the estimates of the specific biomass growth rate and the specific product synthesis rate, one can obtain valuable information concerning the $\pi(\mu)$ relationship.

Furthermore, specific reaction rate estimates may be used as a feedback signal in closed-loop control systems, allowing for more accurate control of these quantities.

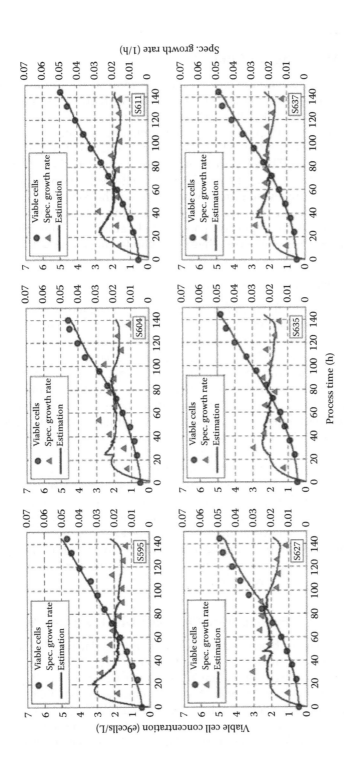

FIGURE 5.11

Biomass and specific growth-rate estimates of the hybrid model compared with the experimental data. (Reprinted with permission from Aehle, M. et al., *Cytotechnology* 62 (5), 413–422, 2010.)

5.4 Hybrid Modeling for Control

Closed-loop control systems are used to keep the controlled bioprocess variables on the predefined set-point profiles, despite the disturbances affecting the real process. When a deviation of the process variables from their predefined trajectories occurs, corrective actions must be actuated. These can be carried out more efficiently when the relevant dynamic features of the process are considered. The advanced closed-loop controllers should involve models for state estimation and control actions: state estimation algorithms are necessary to accurately determine whether or not the controlled variable follows the predefined path, and model-based control algorithms should determine the necessary actions of manipulated variables to eliminate the detected errors. For these purposes, hybrid methods can be utilized.

5.4.1 Requirements for a Process Model

In a closed-loop control, the requirements for a process model are different as compared to models for open-loop process optimization where model generalization and prediction capacities are required for a wide range of variations of process variables. Closed-loop controller actions are based on accurate local predictions with a short time horizon, since quick reactions of the controllers to the disturbances are required. Within short time periods, the controller must bring the process back onto its predefined path. Consequently, the important details of the process dynamics must be known in order to compute precise controller actions to eliminate the identified errors. To achieve this, the process model for controller design must be formulated accordingly.

Different advanced control techniques may be classified into different categories based on the amount of a priori knowledge they use about the process. Figure 5.12 shows some possible variants of process controllers, the applicability of which depends on the knowledge about process. An improvement of the quality of the on-line control can often be achieved using different combinations of control algorithms (hybrid controllers). In the following discussion, the developed on-line control algorithms using hybrid methods are briefly reviewed.

5.4.2 Combination of Feed-Forward Controller with Fuzzy Rule System

The biosynthesis of many biochemical products is closely related to the particular specific growth rate μ; therefore, the careful control of μ-profile is an important task in bioengineering. As an example, a process where the specific growth rate μ of the microorganisms is to be controlled to some desired set point, taking the substrate feeding rate F (fed-batch *S. cerevisiae* cultivation in a bioreactor) as the manipulated variable is considered. This can be achieved by means of a feed-forward/feedback-control scheme, where $F = F_F + F_B$ is computed from

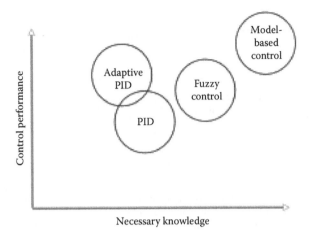

FIGURE 5.12
The performance of closed-loop controllers depends on the amount of knowledge about the process used for control.

two components: the feed-forward rate F_F and the feedback rate F_B. The feedback rate F_B is used to correct the feed-forward rate F_F for deviations of the process state from the state assumed for the correctly running process. The correction may be determined from the deviations of the different measurement values from the desired profiles. From the state variables X, S delivered by the state estimation algorithm, the feed-forward rate F_F is determined by

$$F_F = \frac{\dfrac{\mu_{set}}{Y_{XS}}}{S_{inp} - S} XV \qquad (5.2)$$

where:
  μ_{set} is the desired specific growth rate, [1/h]
  Y_{XS} is the biomass yield, [g/g]
  S_{inp} is the substrate concentration in the feed, [g/l]
  S is the actual substrate concentration in cultivation medium, [g/l]
  X is the biomass concentration, [g/l]
  V is the actual culture volume, [l]

It is essential to stress that for a typical biochemical process (here yeast cultivation):

1. The feed rate profile F_F (t) strongly depends on the set specific growth rate.
2. The process shows completely different behavior for different specific growth-rate profiles.

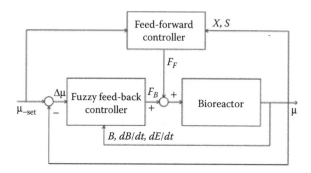

FIGURE 5.13
Hybrid structure of the controller used to control the specific growth rate μ, (1/h), for lab-scale bioreactor, where X, S, and F are biomass, substrate, and ethanol concentrations, (g/L), respectively; B is cumulative base solution amount, (g); dB/dt is base solution consumption rate, (g/h); dE/dt is ethanol production/consumption rate, (g/h); and FB and FF are the feedback and feed-forward control actions, (g/h), respectively.

These two facts explain that the feedback correction term is necessary in adapting the feed rate. When small differences between the expected behavior and the observed one arise, the process may develop in an unexpected way. There are many different ways to construct a controller for the correction term F_B. Here a simple and robust fuzzy expert controller was developed that uses the operators' experience about the process behavior in different process situations. For basic process phases, an appropriate correction F_B was formulated by a set of if/then rules. The general structure of the controller is shown in Figure 5.13.

The control quality was investigated in different lab-scale cultivations. The tests showed a good controller performance (mean absolute error, MAE) between the set point and actual specific growth rate was MAE <0.02 [1/h] during the performed cultivation experiments.

5.4.3 Hybrid Methods and PID Controller

For the past few decades, control theory has attempted to develop exact solutions for well-defined problems. Less research has been dedicated to understanding the heuristics used by experts in industry. At present there is still a large class of problems in control that are solved by heuristics developed by practicing engineers over the years (e.g., controllers are often tuned using rules of thumb).

Recently, knowledge-based control has become an important approach toward the realization of intelligent control that aims to incorporate different sources of information into control systems. In the design of a Proportional-Integral-Derivative (PID) controller it is desirable to incorporate the expert knowledge of design engineers so that the controller can make decisions on the choice of control algorithms and provide diagnoses on the effectiveness of the control system. Based on heuristic knowledge, it can also decide how to tune the PID controller. An efficient alternative is a fuzzy rule system that supervises the process and identifies whether an adaptation of the controller algorithm/parameters is necessary or not.

In order to achieve the advantage with respect to conventional adaptive control, it is necessary to utilize the measurements that are most sensitive to the process dynamics.

5.4.4 Hybrid Models Containing Fuzzy Rule Systems

In recent decades, fuzzy systems emerged as a very convenient and powerful way of processing heuristically based rules of thumb developed by process experts to computation algorithms for the incorporation of fuzzy rule systems into the hybrid modeling environment. Therefore it is an advantage to develop computation algorithms for the incorporation of fuzzy rule systems into the hybrid modeling environment. The advantages of fuzzy rule systems from the point of view of process representation are that a priori knowledge about the process can be used directly in process supervision and prediction software and that these models remain fairly transparent. The powerful tool of ANNs does not yet provide a comparable transparency. However, it would be of considerable advantage to find a way of making use of a priori knowledge in ANNs, and it would be desirable to relate this knowledge to the network structure and its weights (Brown et al. 1994; Wang 1996; Negnevitsky 2004). A promising method in this respect is the technique of *fuzzy artificial neural networks* (Fuzzy ANN), where the expert knowledge about the dynamics of the system, formulated in terms of fuzzy rules, is mapped into the structure of an artificial neural network. An example is displayed in Figure 5.14. In such a way, the weights in the ANN relating the input variables of the process to its output variables can be connected with the physical process variables. The initial values of the weights in these networks can be estimated from the experts' experience. Such fuzzy ANN can then be trained on the available process data in the same way as would a conventional neural network. The immediate result of the training is improved membership functions—terms of the fuzzy variables from the rules formulated by the expert.

There are many advantages of the fuzzy ANN as compared with conventional networks:

- The result of the training can easily be inspected by the process expert.
- The computational effort to train the ANN is considerably reduced (as compared with a conventional network trained on the same data set), since training starts with better initial values and since it already contains a great deal of knowledge that otherwise would have to be learned in a time-consuming way from the data.

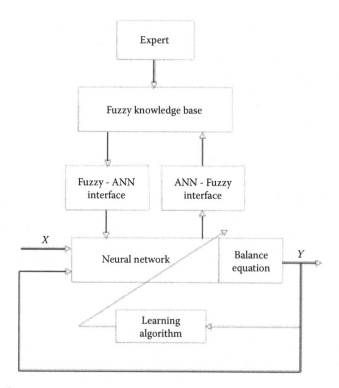

FIGURE 5.14
General scheme of the procedure for arriving at the fuzzy artificial neural network.

- Such networks are also qualitatively improved. Due to the use of real process knowledge, they were shown to provide better extrapolation properties (Simutis et al. 1995). During the test phase of the trained network, it is also easy to detect whether or not the expert considered the entire input data space. If there are gaps, additional rules can be added to the knowledge base without difficulty.

- The tuning of the membership functions in the rule system during network training can be controlled by introducing penalty functions based on heuristical experience. In this way, the experts are able to weight their own statements. In the case that they are sure that their statement is correct, no changes will be allowed by the training procedure.

Obviously, a lot of concrete a priori knowledge is required to formulate the rules—in particular in multivariable processes. Hence, the fuzzy ANN approach cannot be suggested for every application. However, when sufficient a priori knowledge is available that can be formulated in a compact set of fuzzy rules, then the fuzzy ANN provides a stable way of coping with many disadvantages of pure neural network approaches (Simutis et al. 1995).

In particular, fuzzy ANNs provide a transparent way of integrating the background knowledge needed to bridge the gaps in the training data sets leading to the instabilities of the pure neural networks.

5.4.5 Construction Details of Fuzzy ANNs

The main objective in the construction of fuzzy ANNs is to convert a set of fuzzy rules into an appropriate structure, which can be regarded and processed as a feed-forward artificial neural net. The resulting ANN can then be trained on an available data set from the process to be supervised. During the training process, the initial set of fuzzy rules, and specifically the membership functions of the fuzzy variables will be largely improved (Simutis et al. 1995).

Thus, the starting point of the construction is a representation of process input/output relationships by means of fuzzy if/then rules, formulated by process experts.

The l-th of M-rules may be written as

$$R^{(l)} : \underline{IF} \rightarrow x_1 \text{ is } F_1^l \text{ and... and } x_n \text{ is } F_n^l \rightarrow \underline{THEN} \rightarrow y \text{ is } G^l$$

where F_i^l are the fuzzy sets (terms) of the fuzzy input vector $x = [x_1, x_2,..., x_n]$. For the sake of simplicity, the rule system with a single fuzzy output variable y only is considered as an example. Its terms are denoted by G^l.

It is convenient to define the membership functions $\mu_{F_i^l}$ of the terms F_i^l of the fuzzy input variables x_i by Gaussian bell-shaped curves:

$$\mu_{F_i^l}(x_i) = \exp\frac{-(x_i - r_i^l)^2}{2((\sigma_i^l)^2 + \varepsilon)} \tag{5.3}$$

The membership functions of the terms of the output variable y are also described by Gaussian-shape membership functions:

$$\mu_{G^l}(y) = \exp\frac{-(y - s^l)^2}{2((\delta^l)^2 + \varepsilon)} \tag{5.4}$$

The membership functions are thus characterized by two parameters each: the means of r_i^l and s^l and the widths σ_i^l and δ^l. It should be mentioned that the membership functions are normalized to a unit maximum. In order to process the set of fuzzy rules, an appropriate inference mechanism and a defuzzification technique must be chosen. In the case of the training procedure to be applied to the system of rules, the following product inference rule proved to be convenient (Simutis et al. 1995):

$$\mu_{F_1^l F_2^l .... F_n^l} \rightarrow G^l(x, y) = \mu_{F_1^l}(x_1)\, \mu_{F_2^l}(x_2).... \mu_{F_n^l}(x_n)\mu_{G^l}(y) \tag{5.5}$$

For a specific input vector x, this is a function of the output variable y, representing the action proposed by the rule l in terms of a modified membership function.

As an improved defuzzification technique, the *modified center average technique* has been applied (Simutis et al. 1995). It computes the real output value y by

$$
y = \frac{\displaystyle\sum_{l=1}^{M} s^l \left[\prod_{i=1}^{n} \mu_{F_i^l}(x_i) \right] / \left[(\delta^l)^2 + \varepsilon \right]}{\displaystyle\sum_{l=1}^{M} \left[\prod_{i=1}^{n} \mu_{F_i^l}(x_i) \right] / \left[(\delta^l)^2 + \varepsilon \right]} \tag{5.6}
$$

This expression can be interpreted as a weighted average of the means s^l of the individual modified membership functions. The weights consider the degrees to which the conditional part of the individual rules is met by the input vector x and the accuracy by which the expert has formulated the necessary actions. This accuracy is characterized by the widths δ^l of the membership functions defining the terms G^l of the fuzzy output variable y that appears in the action part of rule l.

When the corresponding term (for example, *medium y*) is considered to have a narrow width (i.e., when only values within a narrow interval of the variable y can be regarded as *medium y values*), its mean value s^l should have a larger influence on the final result as if the corresponding interval were wide, indicating that the particular variable value is not so important. Hence, the mean s^l of that membership function will be weighted proportionally to the inverse of the width δ^l. In order to achieve numerical stability, a small quantity ε is added to this part of the weight factor. Experience has shown that ε should be set between 0.001 and 0.01. The feed-forward procedure used to process the M-rules R(l) is as follows. The degree z^l to which the n conditions formulated in Rule l are satisfied for the concrete input vector x is first computed by

$$
z^l = \prod_{i=1}^{n} \exp \frac{-(x_i - r_i^l)^2}{2((\sigma_i^l)^2 + \varepsilon)} \tag{5.7}
$$

The maximum of all z^l (from all rules) measures how completely the expert had considered the input value profiles that the process may assume in practice. In other words, by this value, the software can recognize whether or not the process may run into states that were not considered appropriately by the expert. In situations with a too-low max(z^l), it cannot be expected that the model will work reliably. This property of detecting gaps in the input statements of the rule system or in the membership functions of the fuzzy variables can be used on-line; in other words, it can be used to continuously monitor network reliability. From all these z^l, the answer f of the entire system of fuzzy rules is computed in two steps: In the first of these steps, the weighted sum a of the means s^l of the output functions of the l rules are determined.

$$a = \sum_{l=1}^{M} \frac{z^l s^l}{(\delta^l)^2 + \varepsilon} \tag{5.8}$$

In the second of these steps, the sum b of the weights used is computed

$$b = \sum_{l=1}^{M} \frac{z^l}{(\delta^l)^2 + \varepsilon} \tag{5.9}$$

From both, the response of the rule system to x is computed as the weighted average f simply by

$$f = \frac{a}{b} \tag{5.10}$$

This procedure, as shown in Figure 5.15, can be interpreted as a generalized feed-forward artificial neural network and trained using the classical

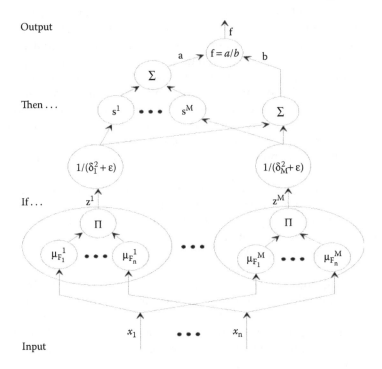

FIGURE 5.15
Scheme of the transformation of the fuzzy rule system to a feed-forward artificial neural network.

back-propagation technique. The parameters to be optimized during the training procedure are the parameters characterizing the individual membership functions

$$w = [r_i^l, s^l, \sigma_i^l, \delta^l], \ i = 1..n \tag{5.11}$$

There are two main differences as compared with conventional ANNs. The first is that the connections reflect the structure of the rules within the fuzzy knowledge base, and hence not all the nodes are interconnected. The second is that the nodes in the different layers perform different calculations.

The convergence of the training algorithm can be improved using the so-called *batching technique*, where corrections to the membership function of the fuzzy variables are not made after each step, but instead after processing a predefined number of experimental data only. This batching algorithm is used during the primary training, while the direct training described in detail was used for the on-line adaptation of the membership functions during the real-time applications of the fuzzy ANN (Simutis et al. 1995).

5.4.6 Hybrid Methods for Imitating Process Control Experts

Experienced, skilled bioprocess operators are able to control their process manually, and when significant deviations from the predefined set-point profiles of process variables appear, the operators can efficiently change the manipulated variables to eliminate these deviations. By designing the advanced automatic control systems, the actions of skilled process operators can be analyzed and transferred into software systems that are capable of imitating the control actions of skilled human operators (Werbos 1992; Bakshi and Stephanopoulos 1996). Figure 5.16 shows the structure of such a hybrid procedure for imitation of the bioprocess control experts. The procedure should imitate humans' ability to recognize various bioprocess situations and to respond appropriately with suitable control actions. Fuzzy reasoning and fuzzy neural networks are the tools that can be involved for the creation of such a procedure. Obviously, the software imitating operators must be developed by process engineers who know how to operate the analyzed bioprocess. The details of an expert imitation procedure can be realized in the following way:

- In the beginning of the imitation procedure, the operator is observed while controlling the process. The actions, performed in different bioprocess situations based on process trends, are converted into fuzzy rules. The resulting fuzzy rule control system can be considered as the initial approach to an expert system imitating control actions of a process operator.
- Since the process operator makes decisions based on shorter or longer trends or shapes of the measured signals, the imitating control

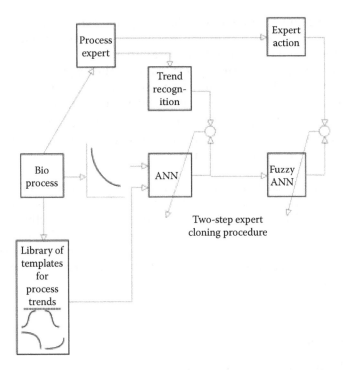

FIGURE 5.16
Use of fuzzy-ANN-based procedures for the bioprocess expert imitation. (Reprinted with permission from Lübbert, A. and R. Simutis, Advances in modeling for bioprocess supervision and control, in *Bioseparation and Bioprocessing: Biochromatography, Membrane Separations, Modeling, Validation*, G. Subramanian (Ed.), Wiley-VCH Verlag GmbH, Weinheim, Germany, pp. 411–461.)

system should also have the ability to recognize and classify the bioprocess trends, which are reflected in measurement signals. For that task, a special library of possible bioprocess trends must be prepared.

- During the cultivation, measurements are collected and trends of various process variables are generated. Then an ANN-based recognizer is employed to identify typical process trends. The information about the process trends is supplied to the fuzzy rule control system, which forms the appropriate control actions.

- In the next steps, this first fuzzy rule control system can be further tuned. This can be carried out by running the control system imitator in parallel to the controlling actions of the operator. The fuzzy rule system can be improved through the deviations between the actions of the human operator and the actions proposed by the imitator control system. This can be carried out by extending the number of rules, by improving existing rules, or by changing the membership functions of the fuzzy variables used.

- After a number of iterations, the fuzzy rule control system cannot learn much more from the process operator. At that point, there is another way to improve the expert imitator: for example, by using evolutionary programming techniques. The approach is to first define a quantitative performance measure for the control and to evolutionarily improve the imitator during real experiments.

- Finally, after the fuzzy rule control system is considered to be essentially complete, the control system can be used to directly control the process or to assist the operator.

In order to avoid misunderstandings, it should be noted here that only human beings are considered intelligent enough to understand a complex process with noises and nonstationary behavior and to directly act on their insights. The imitator fuzzy rule control system is not creative in this respect but can perform many real process control tasks; it is less expensive than a human operator, and it performs control tasks in a much more reproducible way than do humans.

5.5 Hybrid Modeling for Fault Analysis

Reproducible biochemical processes are of high importance for the production of bioproducts.

Hence, during the bioprocess cultivation, it is crucial to perform real-time process monitoring, and if measurement device failures occur (sensor faults—e.g., problems with pH and pO_2 probes) or in case of significant deviations of the bioprocess itself (contamination, genetic drift—e.g., plasmid stability problems), it is necessary to introduce appropriate corrective actions (Royce 1993).

In case of measurement device failures, it is necessary to search for possibilities to correct the functioning of the measurement devices, and after detection of the bioprocess deviations, one has to determine the control actions that would allow the process to be brought back to its normal state.

Traditionally, fault analysis is performed by state estimation algorithms and comparison of the current state with the a priori assumptions of the state variable trajectories. Nevertheless, the implementation of such algorithms requires the mathematical models of the process and the measurement system. The development of such models in many cases requires much effort. Recently, hybrid modeling methods that facilitate the detection and elimination of bioprocess and measurement system faults were being implemented for those purposes (Gnoth et al. 2011).

In advanced industrial biochemical processes, at least 20–30 measurement devices supply the information about the bioprocess in real time. For the process operators, it is rather difficult to follow and interpret this information. Principal component analysis algorithms, auto-associative neural networks, deep-learning neural networks, and other methods were recently developed to allow for transforming the n-dimensional measurement space into 2-D or 3-D feature space with minimal information loss. These features may be used for the development of expert systems that inform process operators about the measurement device faults or cultivation process deviations.

The essence of the hybrid fault analysis methods may be briefly characterized as follows:

- Bioprocess variables that are measured during the cultivation process are mapped from the original (m-dimensional) data space onto a lower (f-dimensional) feature space: $f < m$ and then back from the feature space onto the m-dimensional space using PCA methods, auto-associative neural networks, or other machine learning methods. The dimension reduction algorithms must guarantee that the variables in the feature space contain the main information carried by the original measurements.

- By means of a fuzzy expert system and using the original measurement data, reconstructed measurements, and obtained features, the type of fault occurring during the cultivation process (measurement device fault or bioprocess disturbance) is identified.

- Correction of measurement device signals is performed or bioprocess control strategy is updated.

  The illustration of a hybrid fault analysis method is shown in Figure 5.17.

It is important to stress that such a fault analysis scheme should allow for sufficiently accurate identification of the fault type (measurement device fault or bioprocess disturbance). In case of measurement device fault, clear discrepancies appear between the original data of the analyzed measurement device and the corresponding reconstructed data. In case of certain deviations of the entire bioprocess behavior (contamination, genetic drift—e.g., plasmid stability problems), significant differences between the original and the reconstructed variables are observed. In this case, the variables of the feature layer recede from the feature trajectories of a typical bioprocess. Hence, the information may be directly used for the identification of failures and subsequent correction of the measurement device readings or update of the control strategy of the bioprocess.

The simplest tool for process variable compression and feature extraction is principal component analysis. This technique can easily be implemented in industrial bioprocesses, and it has been used in various applications as an effective tool for process fault analysis. Unfortunately, this technique can perform only linear transformations of measured variables. For nonlinear

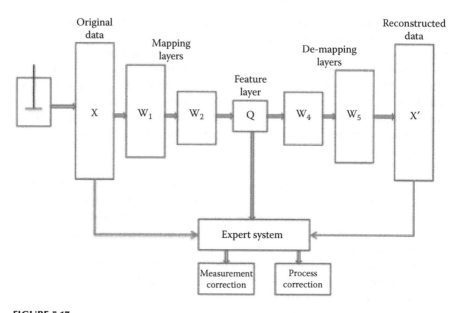

FIGURE 5.17
Hybrid approach for bioprocess fault detection involving an auto-associative neural network and fuzzy expert system.

data transformation, auto-associative neural networks and deep-learning neural networks can be employed. *Auto-associative artificial neural networks* (aANNs) are a special class of feed-forward networks, and they are used to simulate an approximation of the identity mapping between network inputs and outputs using traditional back-propagation algorithms. Recently, for more efficient training of aANN or other deep-learning neural networks with a higher number of mapping/de-mapping layers, some special versions of restricted Boltzmann machine algorithms were used (Hinton et al. 2006). A typical structure of an auto-associative neural network involved for bio-technological process monitoring and fault analysis is shown in Figure 5.18.

The original variables are transformed into the set of features, and then the inverse transformation is applied to form the reconstructed set of variables. The layer containing the features is called a bottleneck layer. It represents a bottleneck in the signal transmission between the set of original variables and its reconstructed version. The network reduces measurement noise by mapping inputs into the bottleneck space, and the residuals between original and reconstructed variables can be used to detect sensor failures. The visualization of the feature space can be very valuable in the process interpretation. The feature time profiles usually differ significantly for *typical* and *untypical* processes. Based on this information, expert rules can be formulated to identify bioprocess faults.

The trained auto-associative network or deep-learning network can also be used to replace unreliable measurement data, once they have been identified

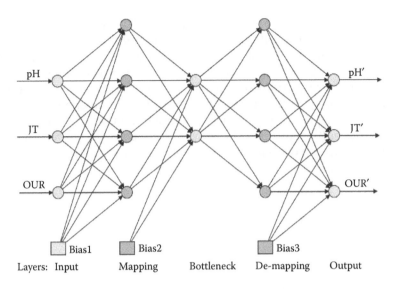

pH JT OUR pH′ JT′ OUR′

Bias1   Bias2   Bias3

Layers: Input   Mapping   Bottleneck   De-mapping   Output

FIGURE 5.18
Example of an auto-associative neural network for biotechnological process monitoring and fault analysis. (Reprinted with permission from Galvanauskas, V. et al., *Bioprocess Biosyst Eng* 26 (6), 393–400, 2004.)

as such. In this case, a fuzzy rule expert system monitors the data, and if the system detects a sensor failure, it gives a signal to an auto-associative network or deep-learning network to replace the unreliable measurement value with estimated values (Gvazdaitis et al. 1994). If the single component X_{fault} of the input vector X is unreliable, then it is possible to search for the best estimate of its value X_{fault}^{est} by systematically varying it within the interval $[X_{fault}^{min}, X_{fault}^{max}]$ of its normal variations until the minimum error E is closely approached. This 1-D search within the known interval can be solved *very* rapidly, since only forward evaluations of the trained auto-associative network or deep-learning network are required. This simple technique does not require many computing resources and can easily be incorporated as an on-line module in the bioprocess control software.

In cases of cell cultures, the processes must be monitored over long cultivation times. In these cases, it is of significant advantage to recognize process faults as early as possible, either to correct the process by changing control variables or to stop the process in order to save time, personnel/equipment costs, and substrate costs. If the associative neural network or deep-learning network has been trained on data from production runs, typical narrow paths of the process in the feature space for correctly running processes can be observed. Experience with these feature trajectories shows that they leave the *typical* tracks observed in correctly running processes, and the trajectories change very sensitively upon process failures.

A typical example of process fault analysis based on feature trajectories is shown in Figure 5.19. Here, feature trajectories are depicted for four industrial protein production processes with recombinant CHO cells. The features were obtained using an auto-associative neural network with a node vector {7 8 2 8 7}, trained on 6 data records. The input vector was composed of the on-line measured signals of carbon production rate, oxygen uptake rate, temperature, glucose concentration, glutamine concentration, pH, and the cultivation time. By means of cross-validation techniques, it was shown that two bottleneck nodes (features) are sufficient to capture the principal dynamics of the process. The trajectories of the features during

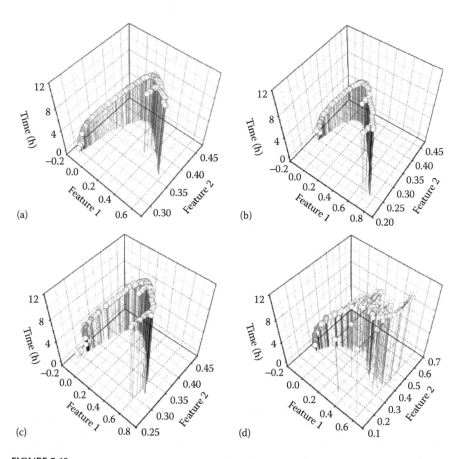

FIGURE 5.19
Application of an aANN to determine process faults during a recombinant CHO cultivation in an industrial erythropoietin production process ([a–c] normal processes, [d] process with faults). (Reprinted with permission from Lübbert, A. and R. Simutis, Advances in modeling for bioprocess supervision and control, in *Bioseparation and Bioprocessing: Biochromatography, Membrane Separations, Modeling, Validation*, G. Subramanian (Ed.), Wiley-VCH Verlag GmbH, Weinheim, Germany, pp. 411–461.)

the cultivation process are depicted in Figure 5.19. In one of the cultivations, an extremely low concentration of recombinant protein was observed. In this case, the features' trajectory (Figure 5.19d) was significantly different from trajectories observed in normal processes (Figure 5.19a, b, and c). Consequently, the trained auto-associative neural network was able to extract important features from the process data, and the obtained features allow for the indication of problems with protein production during fermentation. Based on such typical/untypical process features trajectories, a simple fuzzy rule expert system can be developed for the description and evaluation of the cultivation process behavior. This system allows the early detection of process deviation from the typical path by evaluating standard deviation from the feature trajectories obtained in normal cultivation runs (golden batch). Such system can be easily incorporated into a process monitoring software.

5.5.1 Virtual Plant Concept

New process simulators (*virtual plant* concept) that differ from traditional simulation methods for process design and optimization, in that they run within professional process automation systems using all the original software modules for measurement devices, actuators, and controllers that are finally used by automation systems during their industrial application, are referred to as virtual plants (Kuprijanov et al. 2012). Whereas in a conventional process simulator one models only the dynamic behavior of the cells and their influence on the culture properties, in the virtual plant one needs to additionally introduce the models of the various controllers, sensors, actuators, and other technical hardware modules. It is necessary to fully take into account the dynamic behavior (bioreaction kinetics along with the hardware delays, nonlinearities, etc.) of the whole system (biological and technical) to develop and test a realistic control system. The complexity of the applied models included in such simulators depends on the knowledge of the user about the process and may include submodels that are typical for the hybrid modeling approach discussed in this book.

Since modern process control software includes a high number of basic function blocks and standard control tools, the efforts for virtual plants are less time-consuming than those for conventional comprehensive process simulators, because one can use blocks from a standard library. In conventional process simulators, all these blocks would also have to be created and simulated in order to make sure that the control system is properly designed and tuned.

Virtual plants can be extended in such a way that they can perform the simulations parallel to the action of the process control system during its normal activity. This can be made at the same pace as the process, which allows an on-line comparison between the process and the simulation.

However, advanced virtual plants can make the simulation at a higher speed in order to make a prediction for a given control strategy. In this way, controller details can be developed on-line.

As this requires formulating two control systems—one for the conventional simulator and one for the target automation system—this requires extensive development time (Kuprijanov et al. 2012).

An obvious advantage of a virtual plant that can simulate the process behavior simultaneously to the cultivation is that it can be used as an improved process supervisor and fault analyzer. In its standard application, it runs synchronously to the cultivation, but when significant deviations from the desired path appear, it can run faster in order to see what will happen if the programmed controllers continue to work. Thus, critical process situations can be detected much earlier than with a conventional system. The consequence is that more time is available to cope with the problems.

All modern process control software allows for implementing a virtual plant concept: for example, the process control systems DeltaV (Emerson), 800xA (ABB), TDC (Honeywell), and SIMATIC PCS 7 (Siemens). Kuprijanov et al. (2012) developed and validated a virtual plant within a SIMATIC PCS 7 (Siemens) system and tested it experimentally for various biochemical cultivation processes.

5.6 Concluding Remarks

The application of hybrid models for biochemical process optimization, monitoring, and control may lead to promising improvements of biochemical process quality and safety as compared to the application of pure fundamental, heuristic, or black-box models. The main advantage and driving force of this novel approach is knowledge and information fusion that brings significant added value to modeling quality and process understanding. Hence, this approach could also become an alternative to *big data* technique, which recently became very popular; here, optimization and control solutions are based on pure data analysis and do not take into account the available fundamental knowledge and experience of process experts. Hybrid modeling also addresses the main problems of industrial process monitoring (soft-sensing) and control by means of a knowledge fusion in terms of improving controllability and reproducibility of the biochemical processes as identified in the relevant initiative on process analytical technology (PAT). Nevertheless, to fully utilize the advantages of hybrid modeling, more advanced optimization and control techniques and more sophisticated application software tools need to be developed for the more efficient utilization of its potential.

References

Aehle, M., R. Simutis, and A. Lübbert. 2010. Comparison of viable cell concentration estimation methods for a mammalian cell cultivation process. *Cytotechnology* 62 (5): 413–422.

Åkesson, M., E. Nordberg, K. Karlsson, J. P. Axelsson, P. Hagander, and A. Tocaj. 1999. On-line detection of acetate formation in *Escherichia coli* cultures using dissolved oxygen responses to feed transients. *Biotechnology and Bioengineering* 64 (5): 590–598.

Bakshi, B. R. and G. Stephanopoulos. 1996. Reasoning in time: Modeling, analysis, and pattern recognition of temporal process trends. In *Intelligent Systems in Process Engineering*, G. Stephanopoulos and C. Han (Eds.), pp. 487–546. San Diego, CA: Academic Press.

Brown, M. and C. Haris. 1994. *Neurofuzzy Adaptive Modeling and Control*. New York: Prentice Hall.

Dochain, D. 2008. *Bioprocess Control*. London: John Wiley & Sons.

Galvanauskas, V., R. Simutis, and A. Lübbert. 2004. Hybrid process models for process optimisation, monitoring and control. *Bioprocess and Biosystems Engineering* 26 (6): 393–400.

Gnoth, S., R. Simutis, and A. Lübbert. 2010. Selective expression of the soluble product fraction in *Escherichia coli* cultures employed in recombinant protein production processes. *Applied Microbiology and Biotechnology* 87 (6): 2047–2058.

Gnoth, S., R. Simutis, and A. Lübbert. 2011. Fermentation process supervision and strategies for fail-safe operation: A practical approach. *Engineering in Life Sciences* 11 (1): 94–106.

Gunther, J. C., J. S. Conner, and S. E. Seborg. 2008. PLS pattern matching in design of experiment, batch process data. *Chemometrics and Intelligent Laboratory Systems* 94 (1): 43–50.

Gvazdaitis, G., S. Beil, U. Kreibaum, R. Simutis, I. Havlik, M. Dors, F. Schneider, and A. Lübbert. 1994. Temperature control in fermenters: Application of neural nets and feedback control in breweries. *Journal of the Institute of Brewing* 100: 99–104.

Hinton, G., E. S. Osindero, and Y. W. Teh. 2006. A fast learning algorithm for deep belief nets. *Neural Computation* 18: 1527–1554.

Hjorth, J. S. U. 1994. *Computer Intensive Statistical Methods: Validation Model Selection and Bootstrap*. London, UK: Chapman and Hall.

Jenzsch, M., S. Gnoth, M. Beck, M. Kleinschmidt, R. Simutis, and A. Lübbert. 2006. Open-loop control of the biomass concentration within the growth phase of recombinant protein production processes. *Journal of Biotechnology* 127: 84–94.

Kalman, R. E. 1960. A new approach to linear filtering and prediction problems. *Trans. ASME D: Journal of Basic Engineering* 82: 35–45.

Kuprijanov, A., S. Schaepe, M. Aehle, R. Simutis, and A. Lübbert. 2012. Improving cultivation processes for recombinant protein production. *Bioprocess and Biosystems Engineering* 35 (3): 333–340.

Levišauskas, D., V. Galvanauskas, S. Henrich, K. Wilhelm, N. Volk, and A. Lübbert. 2003. Model-based optimization of viral capsid protein production in fed-batch culture of recombinant *Escherichia coli*. *Bioprocess and Biosystems Engineering* 25 (4): 255–262.

Levišauskas, D., V. Galvanauskas, G. Žunda, and S. Grigiškis. 2004. Model-based optimization of biosurfactant production in fed-batch culture Azotobacter vinelandii. *Biotechnology Letters* 26 (14): 1141–1146.

Lübbert, A. and R. Simutis. 1998. Advances in modeling for bioprocess supervision and control. In *Bioseparation and Bioprocessing: Biochromatography, Membrane Separations, Modeling, Validation*, G. Subramanian (Ed.), pp. 411–461. Weinheim, Germany: Wiley-VCH Verlag GmbH.

Luedeking, R. and E. L. Piret. 1959. A kinetic study of the lactic acid fermentation. Batch process at controlled pH. *Journal of Microbial and Biochemical Technology* 1: 393–412.

Mandenius, C. F. 2004. Recent developments in the monitoring, modeling and control of biological production systems. *Bioprocess and Biosystems Engineering* 26: 347–351.

Negnevitsky, M. 2004. *Artificial Intelligence. A Guide to Intelligent Systems*. New York: Addison Wesley.

Royce, P. N. 1993. A discussion of recent developments in fermentation monitoring and control from a practical perspective. *Critical Reviews in Biotechnology* 13 (2): 117–149.

Schubert, J., R. Simutis, M. Dors, I. Havlik, and A. Lübbert. 1994. Hybrid modeling of yeast production processes-combination of a priori knowledge on different levels of sophistication. *Chemical Engineering & Technology* 17 (1): 10–20.

Shah D., M. Naciri, P. Clee, and M. Al-Rubeai. 2006. Nucleo-Counter: An efficient technique for the determination of cell number and viability in animal cell culture processes. *Cytotechnology* 51: 39–44.

Simutis, R., I. Havlik, and A. Lübbert. 1992. A fuzzy-supported extended Kalman filter: A new approach to state estimation and prediction exemplified by alcohol formation in beer brewing. *Journal of Biotechnology* 24 (3): 211–234.

Simutis, R., I. Havlik, F. Schneider, M. Dors, and A. Lübbert. 1995. Artificial neural networks of improved reliability for industrial process supervision. In *Preprints of the 2nd IFAC Symposium on Modeling and Control of Biotechnical; Processes*, A. Munack and K. Schügerl (Eds.), pp. 59–65. Garmisch-Partenkirchen, Germany.

Simutis, R. and A. Lübbert. 1997. Exploratory analysis of bioprocesses: Using artificial neural network-based methods. *Biotechnology Progress* 13: 479–487.

Simutis, R. and A. Lübbert. 2015. Bioreactor control improves bioprocess performance. *Biotechnology Journal* 10: 1115–1130.

Simutis, R., R. Oliveira, M. Manikowski, S. Feyo de Azevedo, and A. Lübbert. 1997. How to increase the performance of models for process optimization and control. *Journal of Biotechnology* 59: 73–89.

Stephanopoulos, G. and K. Y. San. 1984. Studies on on-line bioreactor identification: Theory. *Biotechnology and Bioengineering* 26: 1176–1188.

Teixeira, A. P., N. Carinhas, J. M. L. Dias, J. P. Crespo, P. M. Alves, M. J. T. Carrondo, and R. Oliveira. 2009. In situ 2D fluorometry and chemometric monitoring of mammalian cell cultures. *Biotechnology and Bioengineering* 102 (4): 1098–1106.

U.S. Food and Drug Administration. 2004. Guidance for industry: PAT—A framework for innovative pharmaceutical manufacturing and quality assurance. http://www.fda.gov/downloads/Drugs/GuidanceComplianceRegulatoryInformation/Guidances/ucm070305.pdf (accessed October 25, 2015).

von Stosch, M., J.-M. Hamelink, and R. Oliveira. 2016. Hybrid modeling as a QbD/PAT tool in process development: An industrial *E. coli* case study. *Bioprocess and Biosystems Engineering* 39 (5): 773–784.

von Stosch, M., R. Oliveira, J. Peres, S. Feyo de Azevedo. 2014a. Hybrid semi-parametric modeling in process systems engineering: Past, present and future. *Computers & Chemical Engineering* 60: 86–101.

von Stosch, M., S. Davy, K. Francois, V. Galvanauskas, J.-M. Hamelink, A. Lübbert, M. Mayer et al. 2014b. Hybrid modeling for quality by design and PAT-benefits and challenges of applications in biopharmaceutical industry. *Journal of Biotechnology* 9: 719–726.

Wan, E. A. and R. van der Merwe. 2001. The unscented Kalman filter. In *Kalman Filtering and Neural Networks*, S. Haykin (Ed.), pp. 221–280. New York: John Wiley & Sons.

Wang, L. X. 1996. *Adaptive Fuzzy Systems and Control*. Englewood Cliffs, NJ: Prentice Hall.

Werbos, P. 1992. Neurocontrol: Where it is going and why it is crucial. In *Intelligent Control Systems*, M. Gupta and N. K. Sinha (Eds.), pp. 791–801. New York: IEEE Press.

6

Hybrid Modeling of Petrochemical Processes

Vladimir Mahalec

CONTENTS

6.1 Introduction

The petrochemical industry produces numerous materials via the transformation of hydrocarbon feedstocks (crude oil or natural gas). A typical process plant (see Figure 6.1) consists of the following:

- Preparation of feedstocks
- Mixing of various feedstocks (this often includes mixing of recycled, unreacted feedstock with the fresh feed)
- Conversion of materials in a reactor
- Separation of the reactor effluent into products and unreacted feedstock
- Recycling of the unreacted feedstock to the inlet of the reactor

Plant design optimizes performance by selecting the best values of design variables (equipment geometry and operating conditions) based on the steady-state models of each piece of equipment in the plant. An equipment model comprises the following:

- Mass balance
- Energy balance
- Performance model specific to the type of equipment

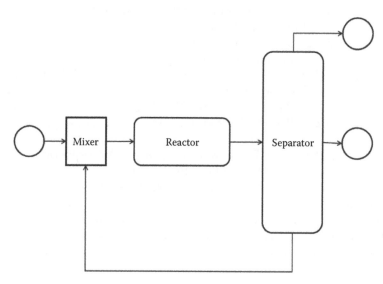

FIGURE 6.1
Typical process system.

- Separation of multi-component mixture into individual products, if applicable
- Material changes due to chemical reactions, if applicable
- Pressure changes
- Heat transfer between phases or compartments of the equipment

During the twentieth century, chemical engineers invested a lot of effort into developing models to predict equipment output based on design variables (e.g., number of trays in a distillation tower, reactor length and diameter, etc.) and operating variables (e.g., reflux ratio in a distillation tower, reactor pressure and temperature). Once detailed equipment design models were developed, engineers started to use them not only for design but also for the optimization of operating conditions in the plants. This trend of using detailed equipment models in plant operations culminated in the real-time optimization of various refining processes and of petrochemical plants. For instance, Dow Chemical Co. has deployed rigorous real-time optimization (PR Newswire, 2001) applications on all of its ethylene plants, while ExxonMobil has implemented real-time optimization on a number of refinery units (Automation.com, 2004).

Although applications of rigorous models for the optimization of plant operations have been successful in many instances, they have also been accompanied by high implementation costs and significant maintenance requirements. A rigorous model of a crude distillation unit can have 20,000 or more nonlinear equations, whereas a complete ethylene plant model may have 100,000 or more equations. Even though detailed rigorous models are used for plant optimization, there is always some discrepancy between the model prediction and the actual plant operation. The development and maintenance of accurate, detailed models of entire plants requires knowledge that is limited to a small number of experts across the world. Hence, a natural question is whether the optimization of existing plants can be carried out by using simpler models that can be developed and maintained by a wide body of engineers.

Hybrid models aim to fill this need for simpler and yet very accurate models that can be used for the monitoring and optimization of operating conditions in process plants. As highlighted in Chapter 2 and shown in Figure 6.2, a hybrid model can predict deviations between the actual process behavior and the predictions from the first-principles models (parallel structure). Such an approach is useful when the process is highly nonlinear and the first-principles model can represent different regions of the process reasonably well but is not sufficiently accurate. Another approach is to use the empirical part of the hybrid model to predict some parameters in the first-principles model (serial structure)—for instance, a heat transfer coefficient in some complex system. The third possibility is to simultaneously solve both the empirical and the first-principles part of the hybrid model

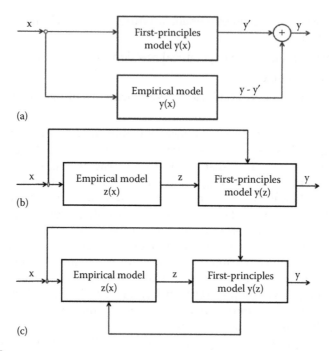

FIGURE 6.2
Structures of a hybrid model: (a) Parallel structure; (b) Serial structure; (c) Simultaneous structure.

(simultaneous structure). For example, the empirical part may predict the fraction of the heavy key in the distillate and the fraction of the light key in the bottoms of a binary distillation tower from energy supplied in the reboiler and the reflux while the material balances are used to compute composition of the streams.

Various forms of equations that describe equipment performance can be chosen. Early efforts to develop rigorous equipment models started with semiempirical equations. An example is Fenske's equation for the separation of a binary mixture in a distillation tower, which uses relative volatility, product purities, and the minimum number of trays. Since such an equation is too simple to predict the performance of an actual distillation tower, its accuracy can be increased either by adding more first-principles-based equations or by estimating some parameters in the semiempirical model via regression. In general, hybrid models (see Figure 6.2) use empirical relationships to do the following:

- Predict (some of) the parameters in the simplified first principles equations (serial structure)
- Predict the discrepancy between the simplified first-principles model and the actual equipment (parallel structure)

- Predict some of the state variables that appear in the simplified first-principle model (simultaneous structure)

The most frequently encountered types of empirical models include these:

- Artificial neural networks (ANN)
- Projection to latent structures (PLS), which is often also called partial least squares
- Support vector regression (SVR)

There are many references available for these methods. For instance, an introductory explanation of neural networks is available as a tutorial at Stanford's UFLDL site, while Kourti and MacGregor (1995) provide an overview of PLS and related methods. An advantage of an ANN is that it can model practically any type of nonlinearity; its disadvantage is that it requires a large set of experimental data in order to train the network properly. It should be noted that prior to the development of an ANN data normalization, removal of redundant and outlier information should be performed to improve the probability of good neural network performance. Even after these preparations, a neural network model may have undesired properties, as illustrated by Szegedy et al. (2014).

On the other hand, PLS and SVR require a much smaller set of experimental data than does ANN, and they can deal with data sets where many variables are cross-correlated. Prior to the development of a PLS model principal component (Shlens, 2014) analysis is used to determine which variables are closely correlated. A disadvantage of linear PLS and SVR is that they cannot represent nonlinear phenomena well. Since plant data as a rule include laboratory analysis of the materials, and since such analysis is available only on some days of the week, the number of available data points from a plant is limited. In addition, on-line plant measurements are highly cross-correlated. These characteristics of plant data make PLS and SVR methods the most suitable for development of empirical models if the underlying phenomena can be represented by linear relationships.

Process plant operating conditions during application of the model will invariably be different from the data used for model development. How can validity of the model be ascertained in the region where the plant is at any point in time? Courrieu (1994) introduced algorithms for estimating the domain of validity of neural network models. Kahrs and Marquardt (2007) generalized that work and proposed two criteria:

1. *Convex hull criterion:* The model input data must be in the convex hull of the data used for model training. This criterion must be used with caution. If most training data are clustered in one region, and a small number of points are distant from it, the model may not be valid throughout the entire convex hull.

2. *Confidence interval criterion*: This criterion takes into account the actual data distribution and the influence of measurement noise on the hybrid model prediction.

As a simple rule of thumb, data used for development of a hybrid model need to cover the expected region of the model application (convex hull criterion) while ensuring that the representative data sets cover all parts of the region.

6.2 Computation of Mass and Energy Balances

Chemical processes convert one material (a chemical compound) to another through chemical reactions. The modeling of changes via chemical reactions requires that the amounts be expressed in moles, not just in mass units. On the other hand, measurements in the plants are as a rule available in volumetric or mass units, not in moles. In addition, the exact chemical composition of the process streams is frequently not available (e.g., composition of kerosene or gas oil in a refinery, or composition of an ore in a smelting plant). Hence, mass and energy balances will be written in mass or energy/mass or energy units. Moles will be used only when conversion from one chemical species to another is required.

Rigorous models rely on rigorous thermodynamic packages to compute properties of the materials—for example, enthalpy per unit mass or boiling-point temperature. A typical goal is to develop models that represent the actual equipment behavior with such accuracy that the root mean square error of the prediction is less than an error acceptable for a specific application of the model. In process plants, measurement errors are typically around 0.1%–1%. Hence, it is reasonable to set a goal of having the root mean square error of the model prediction to be up to 1% relative to the actually measured value of a variable. The same level of accuracy is achieved by, for example, rigorous models of crude distillation unit when applying a model based on rigorous tray-to-tray computation of distillation performance. Hence, approximations of bulk physical properties can be used in order to simplify model calculations and still achieve the target accuracy.

In this chapter, we present development of the models that represent steady-state operation. Hence, the mass balance equation is given by Equation 6.1:

$$\sum_{i=1}^{N\_in} F(i) - \sum_{j=1}^{N\_out} F(j) \tag{6.1}$$

where:
N_in is the number of inlet streams
N_out is the number of outlet streams
$F(i)$, $F(j)$ (*mass/time*) are inlet and outlet streams, respectively

Energy balance for equipment is described by Equation 6.2:

$$\sum_{i=1}^{N\_in} H(i)F(i) + \sum_{K=1}^{NQ\_in} Q(k) - \sum_{j=1}^{N\_out} H(j)F(j) - \sum_{l=1}^{NQ\_out} Q(l) = 0 \qquad (6.2)$$

where:
  $H(i)$, $H(j)$ (*energy/mass*) are inlet and outlet streams enthalpies, respectively
  NQ_in is the number of inlet energy streams
  NQ_out is the number of outlet energy streams
  $Q(k)$, $Q(l)$ (*energy/time*) are the enthalpies of inlet and outlet streams, respectively

Since the hybrid model will be used to represent operation of existing equipment, there must be one or more operating states for which the values of the variables are known. Assuming that for a base case the operating conditions are known—for example, properties of the streams (temperatures, enthalpies, etc.)—the stream enthalpies at some other conditions can be calculated by a linear approximation as shown in Equation 6.3:

$$H(i) = H^0(i) + C_p^0(T - T^0) \qquad (6.3)$$

where:
  $H(i)$, $H^0(i)$ (*energy/mass*) are enthalpies at the current and base case conditions, respectively
  C_p^0 (*energy/mass/deg*) is specific heat capacity at temperature T^0 and pressure P^0

If a process stream contains more than one phase—for example, vapor and liquid or liquid and solids—then the enthalpy of each phase must be calculated separately and added up to compute the total stream enthalpy. Clearly, in such cases the fraction of each phase must be known.

Computation of enthalpies via Equation 6.3 requires knowledge of the specific heat capacity. It should be noted that the specific heat capacity in (energy/mole/degree) units varies widely from one chemical compound to another. On the other hand, the specific heat capacity in (energy/mass/degree) units is almost constant for compounds that have similar bulk properties. A process stream in petrochemical plants is typically made up of similar chemical compounds that have similar specific heat capacities (energy/mass/degree). For instance, a naphtha stream in a refinery consists of hydrocarbons having five to eight carbon atoms. Even though the compositions of different naphtha streams are different, their specific heat capacities (energy/mass/degree) are very close to each other. Hence, if the composition of a given process stream changes due to variations in operating conditions, changes in its specific heat capacity expressed in (energy/mass/degree) units are insignificant. Similar observations are valid for other types of process plants—for example, cement producing or metals producing plants—since specific heats of naturally occurring solids (sand, crushed rock, cement mortars, and others) are practically the same and vary only slightly with temperature (see *Perry's Chemical Engineers' Handbook*). Therefore, using enthalpy in (energy/mass) units in energy balance equations ensures that only a small relative error is caused by approximate enthalpy calculations, as in Equation 6.3.

If the hybrid model is to be used to predict equipment performance in the neighborhood of the base case operation, then the specific heat capacity can be assumed to remain constant even if the stream temperature or pressure varies somewhat. Therefore, for energy balance purposes a given stream in a process can be considered to have specific heat capacity that does not change with variations in the process conditions. Changes of the stream enthalpy with respect to the base case conditions are calculated based on the changes in the stream temperature.

6.3 Hybrid Models of Petrochemical Reactors

Reactor behavior is in most cases nonlinear with respect to operating variables (pressure temperatures, concentrations). Hence, the semiempirical part of the hybrid models needs to be able to deal properly with nonlinearities. This has led to a wide use of artificial neural networks or, more recently, to an increased use of support vector regression models. Several different approaches to development of hybrid reactor models are illustrated in the following sections.

6.3.1 Approximate Heat Transfer Mechanism: Highly Exothermic Fixed-Bed Reactor

A simplified schematic of a tubular fixed-bed wall-cooled reactor is shown in Figure 6.3. One of its applications is to oxidize benzene to produce maleic anhydride. There are two sets of parameters that need to be known in order to model the reactor: (1) reaction system parameters and (2) heat transfer parameters. If reaction system parameters are known, then the remaining issue is the description of the heat transfer across the bed and across the reactor wall. Accurate prediction of reactor performance requires a rigorous two-dimensional model that considers heat transfer in the radial direction (radial thermal conductivity) and heat transfer across the cooling wall (wall heat transfer coefficient).

Qi et al. (1999) compared predictions from a two-dimensional reactor model with a hybrid reactor model that employed a simple plug-flow reactor model as its mechanistic (first principles) part. It is well known that for highly exothermic reactions, a simple plug-flow model cannot predict accurately the performance of the reactor, due to incorrect heat transfer representation. Qi and coworkers employed a three-layer feed-forward

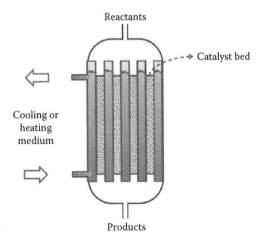

FIGURE 6.3
Tubular fixed-bed reactor.

neural network (see UFLDL tutorial on multilayered neural networks) to estimate the overall heat transfer parameters, which were then used in the plug-flow reactor model. The structure of the hybrid model is shown in Figure 6.4. Both steady-state and dynamic hybrid reactor models were developed. Their predictions were compared (see Table 6.1) to the predictions of the two-dimensional rigorous model and to the experimental data in Figure 6.5 (steady state) and Figure 6.6 (dynamic). Clearly, the

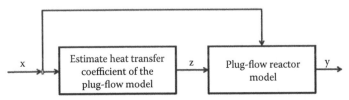

Fixed-bed reactor - hybrid model structure

FIGURE 6.4
Hybrid model structure for the fixed-bed reactor model.

TABLE 6.1

Selectivity Predictions of Hybrid and 2-D Steady-State Models

Experiment number	1	2	3	4	5	6
Experimental data	67.1	68.8	69.6	66.8	68.1	66.2
Hybrid model prediction	67.5	68.9	70.1	67.4	66.8	67.7
2-D model prediction	69.2	68.9	68.6	66.9	69.3	69.5

Source: Data from Haiyu, Q. I. et al., *Chem. Eng. Sci.*, 54, 2521–2526, 1999.

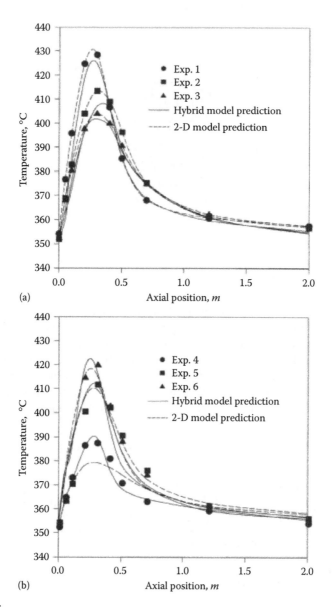

FIGURE 6.5
Steady-state hybrid model predictions: (a) Experiments 1–3; (b) Experiments 4–6. (Reprinted with permission from Haiyu, Q. I. et al., *Chem. Eng. Sci.*, 54, 2521–2526, 1999.)

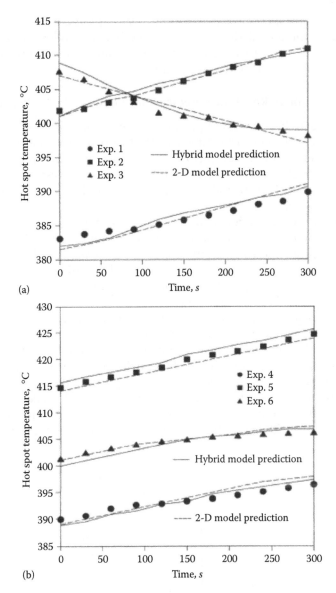

(a)

(b)

FIGURE 6.6

Dynamic hybrid model predictions: (a) Experiments 1–3; (b) Experiments 4–6. (Reprinted with permission from Haiyu, Q. I. et al., *Chem. Eng. Sci.*, 54, 2521–2526, 1999.)

hybrid model predictions are as accurate as the predictions from the two-dimensional model. At the same time, the hybrid model requires much shorter computational times, which make it much more suitable for on-line optimization and control applications.

6.3.2 Approximate Reaction Scheme: Continuously Stirred Reactor

Terephtalic acid (TA) is used to produce polyester, an important material in the petrochemical chain. The industrial production of TA is based on oxidation of p-Xylene (PX) in acetic acid in a continuously stirred reactor (CSTR) in the presence of a catalyst (Figure 6.7). An oxidation process occurs through many steps of a chain of radical reactions containing many radicals and molecular species.

An exact, rigorous model of such a reaction system is very elaborate and includes many intermediate species that are not of interest with respect to commercial production. Hence, instead of using detailed models of the reaction system, a lumped kinetic model that involves only species of interest can be developed. Wang et al. (2005) proposed a kinetic model based on an elaborate radical-chain reaction mechanism. Such a complicated reaction mechanism requires a lot of experimental data that are expensive to obtain. In addition, the mechanism is highly nonlinear, which leads to difficulties in determining the best values of the parameters. Dong and Yan (2014) employed a simplified reaction mechanism model by using a lumped parameter kinetic model. Kinetic parameters in such a model depend on the operating conditions (pressure, temperature, concentrations) since a detailed mechanism is not present in the model. Dong and Yan compared performance of SVR with a radial basis function (RBF) kernel versus three-layer back-propagation

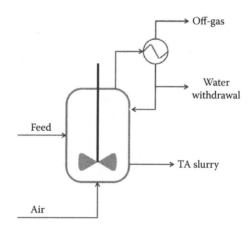

FIGURE 6.7
CSTR reactor for the production of terephtalic acid.

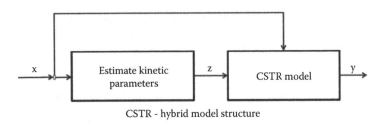

CSTR - hybrid model structure

FIGURE 6.8
Hybrid model of CSTR.

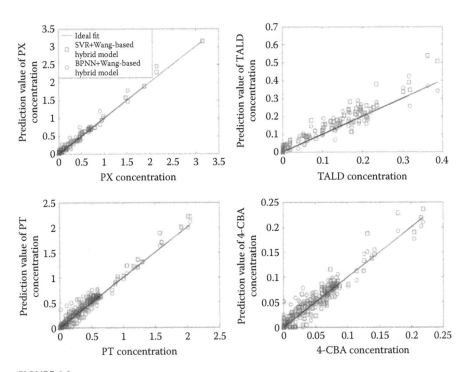

FIGURE 6.9
ANN and SVR hybrid model predictions of PX, TALD, PT, and 4-CBA concentrations. (Reprinted with permission from Yaming, D., Dong, Y., and Yan, X., *Korean J. Chem. Eng.*, 31, 1746–1756, 2014.)

artificial neural networks in estimating kinetic parameters. A simplified representation of the hybrid model is given in Figure 6.8. SVR model predictions were better than ANN model predictions (see Figure 6.9). This is expected, since the number of experimental data points is relatively small. Modeling of this reactor illustrates that with a limited number of available sets of data (which is very often the case in practice) the PLS models (for linear systems) or the SVR models (for nonlinear systems) will perform better than ANN, since ANN models generally require very large sets of data in order to be trained correctly. As seen from Figure 6.9, the hybrid model predicts accurately the concentration of different compounds.

6.3.3 Approximate Reaction Scheme: Fixed-Bed Catalytic Reactor

The hydrogenation of carbon dioxide into methanol in the presence of a catalyst is an important pathway toward reducing CO_2 emissions. The following reactions represent the main changes in the process:

$$CO + 2H_2 \Leftrightarrow CH_3OH \tag{6.4}$$

$$CO_2 + 3H_2 \Leftrightarrow CH_3OH + H_2O \tag{6.5}$$

$$CO_2 + H_2 \Leftrightarrow CO + H_2O \tag{6.6}$$

Research into the kinetics of this hydrogenation has determined that only two of these equations are stoichiometrically independent, which needs to be taken into account when writing the model equations. The reaction is carried out in a Lurgi-type reactor. Reaction mechanisms in the presence of various catalysts are still not fully understood, making rigorous modeling an arduous task. Instead of trying to develop a detailed reactor model (which would have to contain some empirical parameters), Zahedi et al. (2005) developed a hybrid model where a RBF network was used to model the reaction system and to predict concentrations at the outlet of the reactor. The first-principle part of the model included component mass balances, calculation of the pressure drop across a specified length of the reactor bed, energy balance, and equations to compute physical properties of methanol (viscosity, specific heat capacity, heats of reactions) as described next.

Using a differential mixed flow reactor as an approximation, the component mole balances are given by

$$My_i - M^0 y_i^0 = Wr_i \tag{6.7}$$

where:
   M and M^0 (*moles/time*) are the total molar flow at the reactor outlet and inlet, respectively
   y_i is the mole fraction of component i
   r_i (*mole/mass of catalyst/time*) is the rate of formation of component i
   W is mass of the catalyst in the reactor

A segment of the reactor can be approximated by a CSTR reactor operating at constant pressure. Then, the component balance equations can be written as

$$M_i = M_i^0 + W \sum_{i=1}^{NC} r_i \tag{6.8}$$

where $i = 1,...,NC$ are the reaction components (H_2, CO, CO_2, CH_3OH, and H_2O). The component mole fraction can be calculated from Equation 6.9, which can be derived from Equations 6.6 and 6.7:

$$\left[\frac{y_i - y_i^0}{\left(\dfrac{W}{M^0} \right)} \right] - y_i \sum_{i=1}^{NC} r_i = r_i \tag{6.9}$$

Pressure drop across the catalyst bed can be computed from Ergun's equation:

$$\frac{dP}{dz} = -10^{-5} \left[1.75 + \frac{150(1-\varepsilon)}{Re} \right] \left[\frac{(1-\varepsilon)}{\varepsilon^3} \right] \left(u_g^2 \frac{\rho_g}{d_p} \right) \tag{6.10}$$

where:
dP is the pressure drop across dz length of the bed
ε is the void fraction in the bed
u_g, ρ_g, and d_p are gas velocity, density, and packing diameter, respectively

Energy balance in Equation 6.11 can be used to compute the temperature of the gas leaving the bed:

$$\sum_{i=1}^{NC} F_{in}(i) * C_p (T_{in} - T_R) - \sum_{i=1}^{NC} F_{out}(i) * C_p (T_{out} - T_R)$$

$$+ \sum_{i=1}^{N_r} MW_i * \Delta H_r(i) * r_i - UA \left(T - T^{shell} \right) = 0 \tag{6.11}$$

where:

$$\bar{T} = \frac{T_{in} + T_{out}}{2} \tag{6.12}$$

In Equation 6.11, T_R is the reference temperature, C_p is the specific heat capacity, and N_r is the number of reactions (which in this case is equal to 2).

The two equilibrium reaction rates can be obtained from Equation 6.9. Physical properties of methanol are obtained from Equations 6.13 through 6.16:

$$\mu = 67.2 \times 10^{-7} + 0.21875 \times 10^{-7} T \qquad (6.13)$$

$$Cp = 1000(A_i + B_iT + C_iT^2 + D_iT^3) \qquad (6.14)$$

$$-\Delta Hr(1) = 57980 + 35(T - 498.15) \qquad (6.15)$$

$$-\Delta Hr(2) = -39892 + 8(T - 498.15) \qquad (6.16)$$

A model of an industrial reactor was built by dividing the reactor along its length into 30 segments. For each segment, Equations 6.7 through 6.16 were solved. The result was the outlet temperature of that segment, which in turn became the inlet temperature of the next segment. The model predicted concentration of methanol at the reactor outlet, with the error being approximately 5% after 25 days of operation and 6% after 75 days of operation, when compared to the first-principles model. Discrepancy between the real plant and the hybrid model could be addressed by on-line model updating. The hybrid model executed 240 times faster than the first-principles model, which made it much more appropriate for optimization and control in real time.

6.4 Hybrid Models of Simple Distillation Towers

Distillation towers are one of the most-often-used type of equipment in petrochemical and refining plants. The U.S. Department of Energy estimates that distillation consumes 4.8 quadrillion BTU of energy annually in the United States alone. This represents about 40% of energy consumed by petroleum refining and continuous chemical processes (White, 2012). The optimization of distillation operations in real time has the potential to save large amounts of energy. Even though it has become simple to model rigorous distillation towers by tray-to-tray calculations, matching such a model to the plant requires iterative changes to the model (e.g., number of trays in each section, tray efficiencies), and even after all these steps, the final model always has some errors with respect to prediction of the actual equipment performance. Hybrid models represent an alternative to rigorous models as a basis for optimization, since their model parameters are computed to match the actual plant data. Our experience shows that if the structure of the hybrid model is known in advance, then the effort of creating a hybrid model

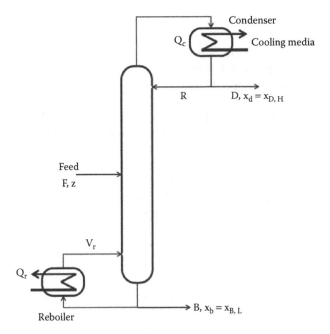

FIGURE 6.10
Two-product distillation tower.

of an actual distillation tower is not any more extensive than the effort to create a rigorous model by configuring commercially available software such as Aspen Plus®. In addition, since a hybrid model has a very small number of equations, it can be implemented easily in digital control system (DCS) systems with minimal requirements of the computational power for their optimization.

Figure 6.10 represents an example of a two-product distillation tower (an example of separation of i-butane from n-butane will be used). A portion of the tower below the feed tray is often called a stripping section, while the portion above the feed tray is called the rectifying section. Most of the time, feed enters the tower approximately in the middle; in other words, the number of theoretical trays in the top and the bottom sections are equal. A hybrid model of the tower consists of mass and energy balances, as well as equations describing separation in the tower.

Overall mass balance:

$$F_F = F_D + F_B \tag{6.17}$$

Component mass balance:

$$f_{Fi} = F_F z_i = F_D x_{Di} + F_B x_{Bi} = f_{Di} + f_{Bi} \tag{6.18}$$

where:

F are flows

z_i and x_i are composition fractions

The subscripts represent F = feed, D = distillate, B = bottoms, and i = 1,...,
number of components (NC)

It should be noted that the component flows that will be predicted from correlations and those that will be computed from material balance equations must be carefully selected. Flows that are small relative to others should be computed from correlations. Large component flows can then be computed by subtracting other (small) flows from, for example, the total flow of the stream. For instance, if the flow of component NC is one of the large component flows that make up the distillate flow, then that flow can be computed as follows:

$$f_{NC} = \left[F_D - \sum_{i=1}^{NC-1} f_{Di} \right] \tag{6.19}$$

where f is a component flow. In Equation 6.19, flows f_{Di}, $i = 1,...,NC-1$ include all small component flows in the distillate. It is not appropriate to compute f_{NC} from Equation 6.19 if f_{NC} is small relative to F_D, since small relative errors in the flows of other components will cause a large error in f_{NC}; in the worst case, f_{NC} can even become negative if Equation 6.19 is used to compute a very small component flow.

Energy balance is given by Equation 6.2. Inlet heat flow to the tower is the energy supplied in the reboiler, while the outlet heat flow is the energy removed in the condenser. In order to predict separation in the tower, approximate equations derived from the first-principles or empirical equations derived from the tower operating data can be used.

6.4.1 Predicting Separation from Semiempirical Equations

Since a rigorous distillation model will not be used, the ability of simplified distillation models to accurately predict changes in the separation of an existing tower that result from changes in the operating variables (reflux, reboiler duty) will be examined. Separation in a distillation tower can be characterized by a separation factor S, which is defined as (Jafarey et al. 1979)

$$S = \frac{\left(x_L/x_H \right)_{top}}{\left(x_L/x_H \right)_{bottom}} \tag{6.20}$$

where:

x is a mole fraction

Indices L and H designate the light key and the heavy key

Note that the separation factor dependence on the product composition is highly nonlinear, which may cause difficulties if the separation factor–based model is used for optimization.

Fenske (1932) derived the overall separation factor for a two-product tower:

$$S = \alpha^N \tag{6.21}$$

where:
  N is the number of stages
  α is the relative volatility

For a tower with N_T stages in the top and N_B stages in the bottom sections, respectively, assuming constant relative volatility, constant molar overflow, and that the feed is optimally located (no pinch zone around the feed), an extended Fenske formula relating internal tower flows, number of stages, and the separation factor is as follows:

$$S \approx \alpha^N \frac{\left(L_{top}/V_{top}\right)^{N_T}}{\left(L_{bottom}/V_{bottom}\right)^{N_B}} \tag{6.22}$$

The separation factor can be calculated from Equation 6.22 by approximating the internal reflux ratio in the top section as a ratio $(R/[D+R])$, while in the bottom section the approximation is $([V_r + B]/V_r)$, where boilup V_r is calculated from the reboiler duty. Comparison of the predictions from Equations 6.20 and 6.22 to the results of a rigorous tower simulation for the butane splitter tower ($N_T = 49$ and $N_B = 50$; $\alpha = 1.221$, $x_{F,nC4} = x_{F,H} = 0.769$) shows that the results obtained from Equation 6.22 often deviate significantly from the rigorous simulation. It should be noted that for a tower with a large number of trays, the calculation via Equation 6.22 is very sensitive to the value of the relative volatility α. The separation factor calculated from Equation 6.22, based on the above-mentioned approximations, is always greater than the actual separation factor calculated from Equation 6.20 based on product compositions from rigorous simulations. Analysis shows that the ratio of the calculated separation factor to the actual separation factor varies with the fraction of the heavy key in the distillate (see Figure 6.11) and with the fraction of the light key in the bottoms product, $x_{B,L}$. Corresponding to each specific value of $x_{B,L}$ is a different straight line:

$$\frac{S_{calc}}{S_{actual}} = A + Bx_{D,H} \tag{6.23}$$

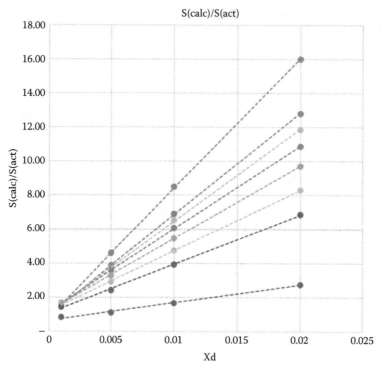

FIGURE 6.11
Ratio of the separation factor calculated from the internal flow estimates to the separation factor computed from actual product compositions.

The slope B and the intercept A in Equation 6.23 depend on the amount of light key in the bottoms product, as shown in Figures 6.12 and 6.13, each of them corresponding to one of the two separate operating regions:

1. Model for distillate compositions from 0.1% to 2% of nC_4 and for bottoms compositions from 0.1% to 2% of iC_4: Slope and intercept in Equation 6.23 depend on the bottoms composition as shown in Figure 6.12a and b. They can be represented by Equations 6.24 and 6.25:

$$A = a_1 x_{B,H}^b \tag{6.24}$$

$$B = \sum_{i=1}^{3} b_i x_{B,H}^{i-1} \tag{6.25}$$

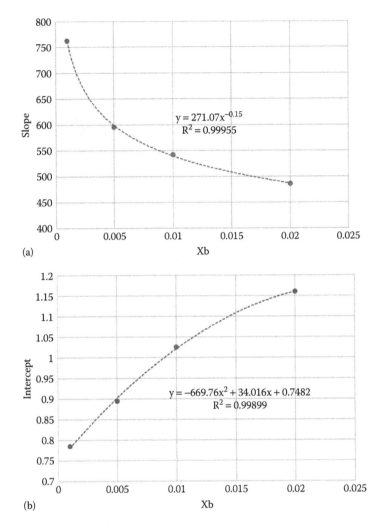

(a)

(b)

FIGURE 6.12
Ratio S(calc)/S(act) for high-purity separations: (a) slope of S_{calc}/S_{actual} versus the amount of light key in the bottoms product and (b) intercept of S_{calc}/S_{actual} versus the amount of light key in the bottoms product.

2. Model for distillate compositions from 1% to 8% of nC_4 and for bottoms compositions from 0.1% to 2% of iC_4. Dependence of the slope and intercept in Equation 6.23 on the bottoms composition is shown in Figures 6.13a and 6.13b and by Equations 6.26 and 6.27:

$$A = a_1 x + a_2 \qquad (6.26)$$

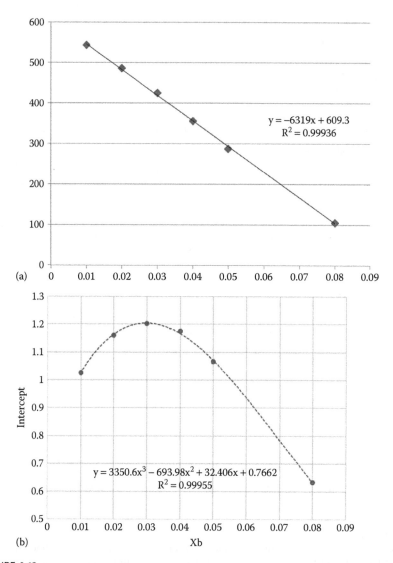

FIGURE 6.13
Ratio $S(calc)/S(act)$ for low purity of bottoms product: (a) slope of S_{calc}/S_{actual} versus the amount of light key in the bottoms product and (b) intercept of S_{calc}/S_{actual} versus the amount of light key in the bottoms product.

$$B = \sum_{i=1}^{4} b_i x_{B,H}^{i-1} \tag{6.27}$$

Equations 6.2 and 6.3 describe the overall energy balances for the tower; Equations 6.17 through 6.20 describe overall material balance and component balances for the tower. Mole fractions of light- and heavy-key components in the distillate and the bottoms streams are given by Equations 6.22 through 6.26. Flows of the components lighter than the light key (or components heavier than the heavy key) are computed by assuming that the entire amounts of such components are transferred from the feed to the distillate (or to the bottoms), as this is common in shortcut computations of distillation towers.

6.4.2 Predicting Separation Factor from Neural Networks

Instead of predicting the separation factor from Equation 6.22 and then finding its relationship to the actual separation factor, one can predict the separation factor directly from the tower operating variables. Safavi et al. (1999) applied a wave-net model (a version of neural networks) to compute the separation factor S. The model contained 64 coefficients. Since the separation factor S can vary over a very large range, they modeled a natural logarithm of S as a function of operating variables. The training data set had about 800 randomly selected data points, while the testing was carried out for approximately 3000 data points. Safavi and colleagues maximized profit by optimizing the hybrid model and compared the results with the optimization of the rigorous model. As shown in Table 6.2, there was a very close agreement between the predictions by the hybrid model and those by the rigorous model. Such high accuracy is not surprising, since there is only a handful of variables that are predicted by the model (e.g., $x_{D,H}$, $x_{B,L}$, S) by selecting the appropriate values for the 64 coefficients. In fact, such a large number of coefficients is analogous to fitting, for example, 5 points with a 6th- or higher-degree polynomial.

Although neural network–based models can predict accurately the behavior of a distillation tower, they have two drawbacks:

1. The training data set must be very large, which makes it expensive to apply them in practice.
2. The structure of ANN models makes them not well suited to be included in models that are used to optimize the operation of a process unit consisting of several pieces of equipment or an entire plant.

TABLE 6.2

Optimization Results from Rigorous and Hybrid Models

Variable	Optimum Rigorous Model	Optimum Hybrid Wave-Net Model
Q(reboiler)	92.525	92.183
$x_{D,H}$	0.72	0.72
$x_{B,L}$	0.0491	0.04887
D	0.617	0.61
B	3.483	3.482
Obj. function	49.95	50.017

Source: Data from Safavi, A. A. et al., *J. Proc. Control*, 9, 125–134, 1999.

Even though neural network models with multiple layers (i.e., deep neural networks) can represent very well the behavior of plant equipment, their structure is highly nonlinear and not amenable to be used as explicit equations in deterministic global optimization algorithms. Hence, optimization of a plant model containing neural network models would have to resort to evolutionary algorithms which currently would lead to very large long execution times and make such applications impractical.

An illuminating discussion of neural networks is presented by Redit (2014).

6.4.3 Predicting Separation from Empirical Equations

Given that versions of semiempirical models that employ the separation factor S are highly nonlinear, the question arises whether simpler models can be derived by using a purely empirical approach. It should be stressed that a purely empirical approach does not mean that all of the knowledge of the process is thrown out the window. On the contrary, all of the available knowledge of the process must be employed to postulate which variables are likely to lead to the simplest form of the model—a linear model. If only statistical methods (e.g., principal component analysis) are employed to identify the most important variables and the understanding of the process is not taken into consideration, the resulting model will be less robust over the entire operating region than if we combine our process knowledge with the statistical methods.

Therefore, to start with the development of an empirical model of separation, distillate impurity versus reflux ratio or bottoms impurity versus boilup ratio is plotted. These curves look similar to a hockey stick. On each side of the bend, the relationship is close to a linear relationship. Hence, the dependence of the product impurity can be approximated by three regions: two linear regions (one for each end of the hockey stick) and one quadratic region corresponding to the bend. Each of these three relationships needs to be modeled separately. Since in an industrial plant the tower usually operates in a relatively narrow region, it means that in practice only one relationship will have to be developed for the target operating region.

In this example, a butane splitter operates in the low-purity region: $x_{B,L} = (0.02, 0.08)$ and $x_{D,H} = (0.02, 0.05)$. Therefore, a model for that region will be presented. It turns out that for the bottom of the tower, the best choice of independent variables is not $([V_r +]/V_r)$; instead of this, it is better to use (V_r/B), since it leads to a linear relationship for bottoms impurity (see Figure 6.14). At the top of the tower, dependence of the distillate impurity on the reflux ratio is linear, with different straight lines for different impurities in the bottom (see Figure 6.15). The slopes of the lines in Figure 6.15 are the

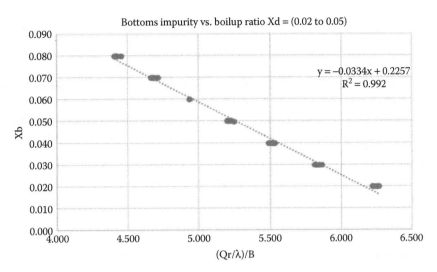

FIGURE 6.14
Bottoms impurity dependence on the boilup ratio.

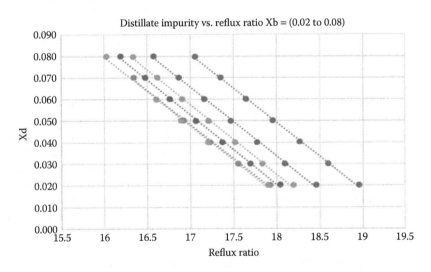

FIGURE 6.15
Dependence of distillate impurity on reflux ratio and bottoms impurity.

same, while the intercepts can be represented by a linear equation. Hence, the following equations represent an empirical model for separation in this butane-splitter tower (for the region of impurities as stated earlier).

Bottoms impurity:

$$x_{B,L} = x_b = c_1 + c_2(Q_r/\lambda)/F_B \qquad (6.28)$$

Distillate impurity:

$$x_{D,H} = x_d = d_1 * (F_R/F_D) + d_2 * x_{B,L} + d_3 \qquad (6.29)$$

Equations 6.28 and 6.29 represent separation in the tower via equations that are much simpler than Equations 6.20 through 6.26. It needs to be stressed that Equations 6.28 and 6.29 should be used only in the region where they have been validated against the plant data or rigorous simulations. Extrapolation outside of the tested region often leads to significant errors.

Once the impurities in the top and the bottom streams are known, the remaining components can be computed via equations from the component material balances.

6.5 Hybrid Models of Complex Distillation Towers

An extreme opposite of a binary mixture is crude oil, which consists of many components. The exact chemical composition of crude oils is not known, since there are hundreds of different crude oils and analysis of their chemical composition would be very costly. Instead of trying to identify the individual chemical components, crude oil is characterized by its boiling-point curve.

Imagine that the chemical composition of the crude oil is known and the constituent components are ordered by their increasing boiling points. These boiling-point values can be used as the *y* axis of a graph. The liquid volume percent (LV%) of each component in the feed is also known from the analysis. Moving from lower to the higher boiling-point components, for each component the cumulative LV% can be calculated for all preceding components plus this component. These values will represent the *x* axis. Such a graph is called a true boiling point (TBP) curve (see Figure 6.16). Since in reality the chemical composition of the crude is not known, the crude distillation curve is measured in a laboratory; such measured curve is discretized in pseudo-components that are separated by, for example, 10°C. These pseudo-components can be characterized in a manner that enables the calculation of their VLE properties, as required for rigorous tower simulation.

Crude oil is separated into several products in a crude distillation unit (CDU). A CDU typically consists of three distillation towers (see Figure 6.17):

1. A preflash tower, which separates very light components (fuel gas, light naphtha) from its bottoms stream, which is the feed to the atmospheric tower

2. An atmospheric distillation tower (atmospheric pipestill, AP), which produces naphtha, kerosene, diesel, and atmospheric gas oil (AGO)

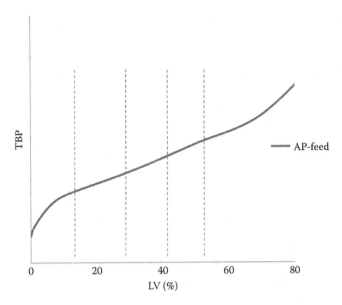

FIGURE 6.16
Crude oil true boiling point (TBP) curve.

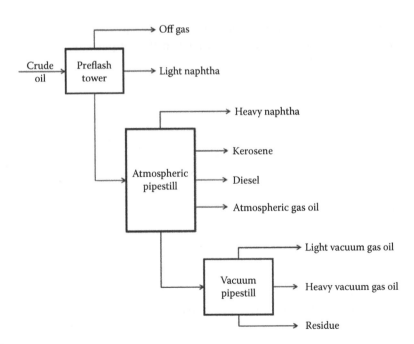

FIGURE 6.17
Typical crude distillation unit.

from its bottoms (reduced crude), which is the feed to the vacuum distillation tower (vacuum pipestill, VP)

3. A vacuum distillation tower (vacuum pipestill), which separates light (LVGO) and heavy (HVGO) vacuum gas oils from pitch

If perfect separation between the products is assumed, then each product TBP curve will be equal to the TBP points on the crude distillation curve (see Figure 6.16). Such perfect separation would require an infinite number of stages, which is impossible. An example of actual separation in an atmospheric pipestill is shown in Figure 6.18. Each product TBP curve is different from the TBP points of the corresponding cut of the crude oil. The front end of each product is lower than the crude TBP, whereas the back end is higher. Product distillation curves in Figure 6.18 were generated from a rigorous model of the sample atmospheric pipestill (see AspenTech "Getting Started Modelling Petroleum Processes" manual).

An accurate approximate model of separation in a distillation tower as complex as an atmospheric pipestill must be constructed, because, together with mass and energy balances, it will constitute a hybrid model of the atmospheric pipestill. If the model is accurate, it will predict product TBP curves that will have very small error when compared to the rigorous model or to the plant data.

There are two aspects of the product TBP curves and related yields:

1. The amount of each product can be adjusted by setting the flow of the corresponding stream; this moves different parts of the feed to different products.

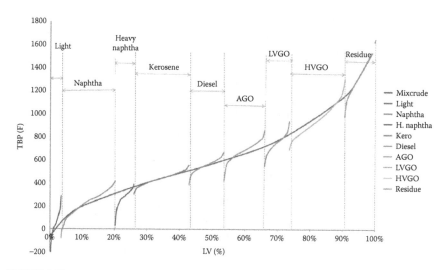

FIGURE 6.18
CDU product distillation curves have overlaps.

By changing the yields of each product (product as a percentage of the feed), we can move the corresponding product cut left or right along the LV% axis of the TBP curve (see Figure 6.16). Moving some product left or right along the LV% axis also changes the boiling points of the material that will be in the product, as a result of changing the product yields. Therefore, changing the product flow (yield) has a major impact on the bulk properties of the product. Since bulk properties are determined mostly by the middle section, it follows that the product yield and the corresponding cut-point temperatures determine the middle section (bulk) of the product TBP curve. The actual slope of the middle section is specific to each AP tower, since the separation also depends on the number of trays in different sections of the tower.

2. For an existing tower, separation between the adjacent products depends on the internal reflux ratios in various parts of the tower, which in turn depend on the operating conditions.

Therefore, the product TBP curves can be predicted in two steps:

1. Compute the middle section of the product TBP curve as a straight line that depends on the product yields, feed TBP curve, and specific gravity. The latter accounts for the fact that different types of crude feed (light or heavy) will separate differently in the CDU (Mahalec and Sanchez 2010).

2. Compute the distance from the TBP curve to the straight line through the middle section; this distance will be computed separately for discrete points on the product TBP curve (see Figure 6.19). The smaller

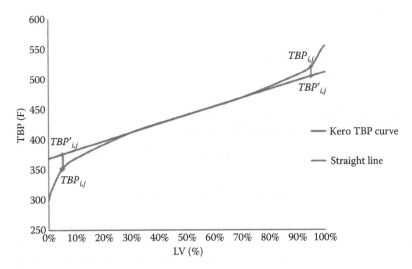

FIGURE 6.19
Computation of end sections of TBP curve of AP product.

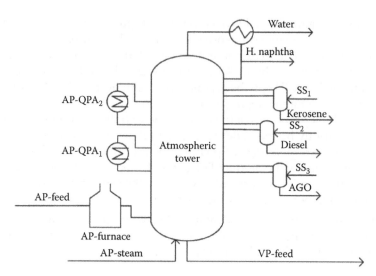

FIGURE 6.20
Atmospheric pipestill configuration.

the distance, the better the separation; changes to this distance are governed by the internal reflux ratios. Since the computation of the internal reflux would require knowledge of the tray temperatures, instead of using the internal reflux, the dependence of the separation on variables that determine the internal reflux (external product flows/yields, heat removal via pumparounds, heat supply to the feed/vapor flow from the feed tray, stripping steam flows) can be exploited.

For the sake of brevity, we will present an abbreviated computation of the product TBP curves for a CDU (see Figure 6.20). Details of the complete model can be found in Fu et al. (2016).

6.5.1 Predicting the Straight Line through the Middle Section of the Product TBP Curve

The cumulative cut width of each product ($CCWP_i$) is defined as

$$CCWP_i = \sum_{k=0}^{i} \frac{F_{p,k}}{F_{feed}}, \, i = 0,1,2,3 \tag{6.30}$$

where $F_{p,k}$ and F_{feed} are the flow rates of product k and of the feed, respectively. The cut-point temperature (CT_i) of each product can be calculated from the feed TBP curve as shown in Figure 6.21. Note that the distillate is labeled as 0,

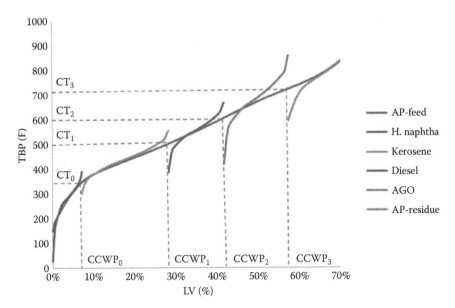

FIGURE 6.21
Definition of product cumulative cut width and cut-point temperature.

so that 1 corresponds to the first sidestream, 2 corresponds to the second sidestream, and so on (counting from the top of the tower).

The straight line through the middle section of individual products is then given by Equations 6.31 through 6.33.

Heavy naphtha (liquid distillate):

$$TBP_{i,j} = a_0(i,j) + a_1(i,j) * \rho_{AP\_feed} + a_2(i,j) * CCWP_i + a_3(i,j) * CT_i \quad (6.31)$$

where:
$i = 0$ (distillate, heavy naphtha)
$j = 50,70$ (since points below 50% do not lie on the straight line)

6.5.1.1 Sidestream Products

$$TBP_{i,j} = a_0(i,j) + a_1(i,j) * \rho_{AP\_feed} + a_2(i,j) * CCWP_i$$
$$+ a_3(i,j) * CT_{i-1} + a_4(i,j) * CT_i \quad (6.32)$$

where:
$i = 1,2,3$ (kerosene, diesel, atmospheric gas oil)
$j = 30,50,70$

6.5.1.2 Bottoms Stream—VP Feed

$$TBP_{i,j} = a_0(i,j) + a_1(i,j) * \rho_{AP\_feed} + a_2(i,j) * CCWP_{i-1} + a_3(i,j) * CT_{i-1}$$
$$+ a_4(i,j) * CT_{AP\_feed,70}$$
(6.33)

where:
  $i = 1$ (bottoms product)
  $j = 30,50,70$

6.5.2 Predicting TBP Curve Front-End and Back-End Deviations from the Straight Line

The deviations from the straight line are defined as

$$TBP_{i,j}^d = TBP_{i,j}' - TBP_{i,j}, i = 1,2,3,4 \text{ (products) and } j = 1,5,10 \text{ } (LV\%) \quad (6.34)$$

$$TBP_{i,j}^d = TBP_{i,j} - TBP_{i,j}', i = 0,1,2,3 \text{ (products) and } j = 90,95,99 \text{ } (LV\%) \quad (6.35)$$

As mentioned earlier, the ends of the product TBP curves vary with changes in operating conditions. At the top of the tower, the internal reflux ratio RR can be approximated by (reflux/[reflux + distillate]): (RR=R/[R+D]). It should be noted that the pumparound duties are as a rule much smaller than the heat removed in the condenser. Therefore, in order not to have a large error in the pumparound duties, these need to be specified, and the condenser duty (corresponding to [R+D]*λ_{vap}) is calculated from the energy balance. Let $y_{AP,i}$ be the yield of the product i (product flow/feed flow); φ the vapor fraction of the feed rising from the feed tray; $\Delta T_{G(i,j)}$ the temperature difference between selected points located at the crude TBP sections corresponding to products i and j; $F_{SS,i}$ the flow of stripping steam to the product i striper; and RR the internal reflux ratio at the top of the tower. Then, the deviations for the front and the back sections are given by Equation 6.34 through 6.37.

6.5.2.1 Back End of Heavy Naphtha/Front End of Kerosene

$$TBP_{i,j}^d = a_0(i,j) + a_1(i,j) * y_{AP,i-1} + a_2(i,j) * y_{AP,i} + \frac{a_3(i,j) * F_{SS,i}}{F_{AP,i}}$$
$$+ a_4(i,j) * \Delta T_{G(i-1,i)} + a_5(i,j) * CCWP_{i-1} + a_6(i,j) * \varphi$$
(6.36)

where:
  $i = 0$ (heavy naphtha)
  $j = 90,95,99$ LV% or $i = 1$ (kerosene) and $j = 1,5,10$ LV%

6.5.2.2 Back End of Kerosene, Diesel, and Atmospheric Gas Oil

$$TBP_{i,j}^d = a_0(i,j) + a_1(i,j) * y_{AP,i} + a_2(i,j) * y_{AP,i+1} + \frac{a_3(i,j) * F_{SS,i+1}}{F_{AP,i+1}}$$

$$+ a_4(i,j) * \Delta T_{G(i,i+1)} + a_5(i,j) * CCWP_i + a_6(i,j) * \varphi \qquad (6.37)$$

where:
 $i = 1,2,3$ (kerosene, diesel, atmospheric gas oil)
 $j = 90,95,99$ LV%

6.5.2.3 Front End of Kerosene, Diesel, Atmospheric Gas Oil, and Bottoms Product

$$TBP_{i,j}^d = a_0(i,j) + a_1(i,j) * y_{AP,i-1} + a_2(i,j) * y_{AP,i} + \frac{a_3(i,j) * F_{SS,i}}{F_{AP,i}}$$

$$+ a_4(i,j) * \Delta T_{G(i-1,i)} + a_5(i,j) * CCWP_{i-1} + a_6(i,j) * \varphi \qquad (6.38)$$

where:
 $i = 2,3,4$ (diesel, atmospheric gas oil, VP feed)
 $j = 1,5,10$ LV%

6.5.2.4 Back End of Bottoms Product—VP Feed

$$TBP_{i,j}^d = a_0(i,j) + a_1(i,j) * CCWP_3 + a_2(i,j) * CT_3 + a_3(i,j) * \varphi \qquad (6.39)$$

where:
 $i = 4$ (VP feed)
 $j = 90,95,99$

Predictions of the product TBP curves computed from the previous equations have been compared to a set of results from the rigorous tray-to-tray model in Aspen Plus. The root mean square prediction error was about 0.25% or less with respect to the results computed from the rigorous model. Considering that the hybrid CDU model has about 160 equations (mostly linear) whereas the rigorous model has 10,000–20,000 nonlinear equations, the accuracy of prediction from the hybrid model is remarkable. Most importantly, the model is simple enough to be included in refinery production planning and scheduling optimization applications.

6.6 Summary

It has been illustrated with several examples how to build hybrid models of chemical reactors and distillation towers—two of the most important types of equipment found in process plants. When building a hybrid model, we must keep in mind these factors:

1. The model needs to be sufficiently accurate to meet the needs of the user. For instance, a 0.25% prediction error from a CDU model that will be used for production planning is an acceptable accuracy, since the data used in production planning have uncertainties much higher than 0.25%.

2. The empirical or semiempirical relationships in the model are valid in the region covered by the training data, and there will always be some prediction errors (small or not so small).

3. Use your knowledge of the process, combined with PCA analysis, to identify which variables are the most important with respect to the system output.

4. As much as possible, use the PLS method to build the model. A plot of PLS model predictions versus observed data readily shows whether the process is linear or not. If the plot is nonlinear, it is frequently possible to correlate the actual plant data with the model predictions. An alternative in such situations is to apply SVR as the model-building method. As an alternative, one can also employ neural networks, while making sure that the network has as simple a structure as possible in order to avoid overfitting.

5. To ensure feasible and meaningful predictions from the model, it is best to predict intensive properties (e.g., the fraction of an inlet component flow that is present in a given product stream, the composition fraction, etc.). In addition, as much as possible, predict those components that are present as relatively low fractions in the product streams; compute the largest component by subtracting other components from the total stream amount (i.e., from material balance). Similarly, specify smaller energy flows (or estimate them from empirical models) and compute the largest energy flow from energy balance.

Finally, let's emphasize that the hybrid models should be used if rigorous first-principles models are unavailable or are unsuitable (e.g., it may take too long to compute the results, or their maintenance may be too complex for the staff at the site).

Acknowledgments

Our work on hybrid models would not have taken place without my graduate students. I would like to acknowledge Asaad Hashim (who worked on the first AP tower hybrid model), Hamza al Shwaragi, Yoel Sanchez (development of AP tower model for real-time optimization), Gang Fu (development of a complete CDU model for planning, scheduling, or RTO), and Fahad al Juhani (hybrid models of two- or three-product distillation towers).

Financial support from Imperial Oil, BP North America, NSERC, Ontario Research Foundation, Saudi Aramco, and McMaster Advanced Control Consortium is gratefully acknowledged.

Nomenclature

C_p	Heat capacity at constant pressure
$CCWP$	Cumulative cut width of a CDU product
CT	Cut-point temperature
F	Mass or volume flow of a stream
F	Mass flow of a component
H	Enthalpy per unit of mass
L	Internal liquid flow in a distillation tower
$LV\%$	Liquid volume percent
M	Molar flow
MW	Molecular weight
P	Pressure
Q	Energy flow
R_i	Rate of formation of component i
RR	Reflux/(reflux + distillate)
S	Separation factor for two-product distillation tower
T	Temperature
V	Internal vapor flow in a distillation tower
W	Mass of catalyst in a reactor
ΔH_r	Heat of reaction
P	Density
φ	Vapor fraction

Indices

AP	Atmospheric pipestill
B	Bottoms
D	Distillate
F	Feed
R	Reflux
SS	Stripping stream

References

al Juhani, F. 2015. Hybrid models of distillation towers. MASc thesis, McMaster University, Hamilton, ON, Canada.

Aspen Technology Inc. 2007. Getting started modeling petroleum processes. Burlington, MA.

Automation.com. 2004. ExxonMobil selects Invensys as provider of real-time optimization software. 2004. http://www.automation.com/automation-news/industry/exxonmobil-selects-invensys-as-provider-of-real-time-optimization-software.

Courrieu, P. 1994. Three algorithms for estimating the domain validity of feedforward neural networks. *Neural Networks* 7 (1): 169–174.

Dong, Y. and X. Yan. 2014. A comparative study of hybrid models combining various kinetic and regression models for p-xylene oxidation. *Korean Journal of Chemical Engineering* 31 (10): 1746–1756.

Fenske, M. R. 1932. Fractionation of straight-run Pennsylvania gasoline, *IEC* 24 (5): 482–485.

Fu, G., Y. Sanchez, and V. Mahalec. 2016. Hybrid model for optimization of crude distillation units. *AIChE Journal* 62: 1065–1078.

Green, D. W., Perry, R. H. 2002. *Perry's Chemical Engineering Handbook*, 8th edition, New York: McGraw-Hill Education.

Jafarey, A. J., M. Douglas, and T. J. McAvoy. 1979. Short-cut techniques for distillation column design and control: 1. Column design. *Industrial & Engineering Chemistry Process Design and Development* 18: 197–202.

Kahrs, O. and W. Marquardt. 2007. The validity domain of hybrid models and its application in process optimization. *Chemical Engineering and Processing* 46: 1054–1066.

Kourti, T. and J. MacGregor. 1995. Process analysis, monitoring and diagnosis, using multivariate projection methods. *Chemometrics and Intelligent Laboratory Systems* 28: 3–21.

Mahalec, V. and Y. Sanchez. 2012. Inferential monitoring and optimization of crude separation units via hybrid models. *Computers & Chemical Engineering* 45: 15–26.

PR Newswire. 2001. Dow Chemical kicks off worldwide implementation of AspenTech production optimization Solutions. 2001. Accessed April 11, 2015, http://www.prnewswire.com/news-releases/dow-chemical-kicks-off-worldwide-implementation-of-aspentech-production-optimization-solutions-74038692.html.

Qi, H., X. G. Zhou, L. H. Liu, and W. K. Yuan. 1991. A hybrid neural network-first principles model for fixed-bed reactor. *Chemical Engineering Science* 54: 2521–2526.

Redit, S. 2014. 10 misconceptions about neural networks. Accessed August 20, 2015. http://www.turingfinance.com/misconceptions-about-neural-networks/.

Safavi, A. A., A. Nooraii, and J. A. Romagnoli. 1999. A hybrid model formulation for a distillation column and the on-line optimization study. *Journal of Process Control* 9: 123–134.

Shlens, J. 2014. A tutorial on principal component analysis. Accessed August 20, 2015. https://arxiv.org/pdf/1404.1100.pdf.

Skogestad, S. and M. Morari. 1987. Shortcut models for distillation columns: I. Steady state behavior, Chapter 9 in PhD diss., *Studies on robust control of distillation columns*, S. Skogestad (Ed.), California Institute of Technology, Pasadena, CA.

Szegedy, C., W. Zaremba, I. Sutskever, J. Bruan, D. Erhan, I. Goodfellow, and R. Fergus. 2014. Intriguing properties of neural networks. Accessed August 20, 2015. https://arxiv.org/abs/1312.6199.

Thompson, M. L. and M. Kramer. 1994. Modeling chemical processes using prior knowledge and neural networks. *AIChE Journal* 40: 1328–1240.

Wang., Q., X. Li, L. Wand, Y. Cheng, and G. Xie. 2005. Kinetics of p-xylene liquid-phase catalytic oxidation to terephtalic acid, *IECR* 44 (2): 261–266.

UFLDL Tutorial: Multi-layered neural networks. Stanford University. Accessed August 20, 2015. http://ufldl.stanford.edu/tutorial/supervised/MultiLayer Neural Networks/.

White, D. C. 2012. Optimize energy use in distillation. *Chemical Engineering Progress* (March): 35–41.

Zahedi, G., A. Elkamel, A. Lohi, A. Jahanmiri, and M. R. Rahimpor. 2005. Hybrid artificial neural network: First principles model formulation for the unsteady state simulation and analysis of a packed bed reactor for CO_2 hydrogenation to methanol. *Chemical Engineering Journal* 115: 113–120.

7

Implementation of Hybrid Neural Models to Predict the Behavior of Food Transformation and Food-Waste Valorization Processes

Stefano Curcio

CONTENTS

7.1 Introduction

Many approaches have been proposed in the literature to predict the behavior of food transformation and food-waste valorization processes. Generally speaking, a theoretical approach, leading to first-principle models, makes use of conservation laws/kinetic equations, which provide an accurate knowledge of the actual system behavior (Andrić et al. 2010). However, an exhaustive analysis of all the complex phenomena occurring in a real process is often difficult to accomplish due to a series of not-completely-understood or difficult-to-be-described phenomena. A significant level of uncertainty actually exists and does not allow one to achieve a rigorous model formalization by proper mathematical relationships (in many cases having real significance). A valid alternative to theoretical modeling is the empirical approach, which leads to a so-called black-box model. A black-box model does not make use of any kinetic or transport equation, which could help determine the mutual relationships between the system inputs and its responses. Among black-box models, artificial neural networks (ANNs) generally provide reliable predictions of real bioprocesses behavior.

ANNs are data-driven models capable of learning from examples, and they are composed of interconnected computational elements, called neurons or nodes, which operate in parallel. Each neuron receives signals from the related units, elaborates these stimuli by a transfer function, and generates an output, which is subsequently transferred to other neurons belonging, in a forward configuration, to a succeeding layer (Reilly et al. 1990). Even if the prediction of each single neuron could be imperfect and bias affected, the outcome of the interconnection(s) among neurons is a reliable computational tool capable of learning from examples and providing accurate predictions even with examples never exploited before (Zhang et al. 1998; Basile et al. 2015). This feature makes ANNs a particularly useful tool when the behavior of complex systems is to be described, since no a priori knowledge of system dynamics is actually required. A neural network model, however, can be rather complicated, since it may require a large number of connections and therefore contain a great number of parameters that need to be estimated. Generally, a larger number of neurons results in a more powerful network, but also in a higher computational effort. In addition, it is worthwhile paying attention to the so-called overfitting, which occurs when a model is extremely complex—when the number of parameters is excessive as compared to the number of observations (see also Chapter 4). A model that has been overfit has poor predictive performance and may overreact to minor fluctuations in the training data (Chen et al. 2016; Estiati et al. 2016; Grahovac et al. 2016).

A reasonable trade-off between theoretical and neural network approaches is represented by hybrid neural modeling, leading to a so-called *gray-box* model capable of good performance in terms of data interpolation and extrapolation (Agarwal 1997). Hybrid neural model (HNM) predictions are given as a combination of both a theoretical and a *pure* neural network approach, together concurring at the obtainment of system responses. As shown in Chapter 2, the main advantage of hybrid neural modeling regards the possibility of describing some well-assessed phenomena by means of a theoretical approach, leaving the analysis of other aspects, difficult to interpret and describe in a fundamental way, to rather simple *cause-effect* models (Simutis et al. 1995; Curcio et al. 2009). When ANNs are exploited in a hybrid model, it is crucial to clearly define which features of the process are to be described by a black-box model and which ones by fundamental equations. Hybrid neural models have been extensively proposed to predict the behavior of either food transformation (Cubeddu et al. 2014; Al-Mahasneh et al. 2016; Neethu et al. 2016) or food-waste valorization processes (Saraceno et al. 2010, 2011).

In this chapter, two case studies are described with the aim of showing that ANNs, combined with a set of theoretical equations, can be used to model two processes that are interesting for the food industry. The first of these case studies deals with the formulation of a hybrid neural model aimed at predicting vegetables drying behavior over a wide range of operating conditions. The proposed model, accounting for the simultaneous transfer of momentum, heat, and mass—both in the drying air and in the food sample—was capable of describing the effect of operating conditions on both microbial population abatement (and, therefore, on bacterial starvation) and color degradation, chosen as a reference quality parameter. It is well known that the exploitation of drastic operating conditions, although improving food safety, may induce major thermal damages, thus determining a significant worsening of dried product quality. In fact, the utilization of mild operating conditions improves the organoleptic properties of dried foods but may not assure a proper decontamination of the final product.

The second of the considered case studies regards the exploitation of a hybrid neural paradigm to model the enzymatic transesterification of waste oil glycerides—the key step for the obtainment of second-generation biodiesel. The possibility of using waste vegetable oils for biodiesel production represents a very interesting alternative to current technologies since it allows waste valorization. In a previous paper (Calabrò et al. 2009), it was shown that transesterification, performed by lipase extracted from *Mucor miehei* and immobilized on an ionic exchange resin, could reliably be described by a Ping-Pong Bi-Bi kinetic mechanism. The obtained theoretical predictions were indeed in good agreement with the experimental results. The kinetic analysis was carried out in order to investigate the effect of two key parameters: the molar ratio of reactants and the ratio between enzyme and substrate fed to a batch bioreactor. In another work (De Paola et al. 2009), a factorial analysis was implemented to account for the mixing rate, the amount of enzyme, and the reactants ratio fed to the bioreactor. The study, based on experimental design, allowed optimizing process performance—maximizing the final conversion of glycerides. However, both of the previous studies were based on some assumptions, the most important of which pertained the existence of a linear relationship determined by analyzing through a fitting procedure the whole set of collected experimental data, between reactants and product concentrations. Starting from the results obtained in Calabrò et al. (2009) and De Paola et al. (2009), it was the intention of this chapter to critically revise the methodology that brought the authors to assume that reactants and product concentrations were related by a linear relationship. This was achieved by more-accurate computational methods, which, based on artificial neural networks, allowed estimating the complex relations existing between process parameters and the true reaction pathway. In particular, it will be shown that a hybrid neural approach can successfully be applied to model biotechnological processes and represents, in the case of biocatalytic transesterification of waste vegetable oils, a powerful tool offering very precise predictions of the actual system behavior over a wide range of process and operating conditions.

7.2 Case Study 1: Convective Drying of Vegetables

7.2.1 Hybrid Neural Model Development

When dry and warm airflows around a moist and cold food sample, two different transport mechanisms simultaneously occur: heat is transferred from air to food; water, instead, is transferred from food to air. Within the solid material, transport of water, as liquid, is due to both gas pressure and capillary pressure gradients, whereas vapor is transferred by pressure and concentration gradients. In addition, heat is transported by conduction and convection.

To develop the theoretical part of the present HNM, it was assumed as follows (Curcio and Aversa 2014): (a) Capillary pressure prevailed over gas pressure, assuming that the hygroscopic material under study was highly unsaturated, as in most drying applications; the transport of liquid water, therefore, occurred essentially by capillary pressure. (b) The term containing the pressure-driven flow in the vapor transfer equation was neglected, and the molecular diffusion was considered the prevailing mechanism. (c) The contribution of convection to heat transport was considered negligible as compared to conduction. (d) Evaporation occurred over the entire food domain and at food outer surfaces. (e) Moisture removal from the surface took place by vapor transport, diffusing into the boundary layer developing in the drying air, and by liquid water transport, evaporating at the outer food surfaces. Both vapor and liquid water were convected away by drying air, whose velocity field is expected to strongly affect the interfacial rates of heat and mass transfer. (f) The continuity of both heat and mass fluxes occurred at the food/air interfaces. (g) All phases—solid, liquid, and gas—were continuous and were in local thermal equilibrium. (h) The vapor pressure was expressed as a function of both temperature and moisture content. (i) Finally, the turbulent momentum transfer, referred to drying air flowing in the drying chamber around the food sample, can be described by the k-ω model.

On the basis of the above-mentioned assumptions, a system of transport equations referred to both drying air and food was obtained to model the behavior of the drying chamber shown in Figure 7.1.

In particular, the unsteady-state mass and energy balance equations referred to the transport of liquid water and of vapor in the food sample were

$$\partial C_w/\partial t + \underline{\nabla} \cdot (-D_w \underline{\nabla} C_w) + \dot{I} = 0 \tag{7.1}$$

$$\partial C_v/\partial t + \underline{\nabla} \cdot (-D_v \underline{\nabla} C_v) - \dot{I} = 0 \tag{7.2}$$

where:
  C_w was the liquid water concentration
  C_v was the vapor concentration

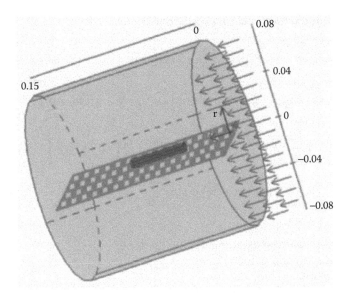

FIGURE 7.1
Schematic representation of the drying chamber.

$\dot{I}$ was the volumetric rate of evaporation
D_w and D_v were the capillary diffusivity of water and the effective diffusion coefficient of vapor in food, respectively

The energy balance in the food material led, according to Fourier's law, to the unsteady-state heat transfer equation

$$\rho_s C_{ps} \frac{\partial T}{\partial t} - \underline{\nabla} \cdot \left(k_{eff} \underline{\nabla} T \right) + \lambda \cdot \dot{I} = 0 \tag{7.3}$$

where:
T was the food temperature
ρ_s was the density of food sample
C_{ps} was its specific heat
k_{eff} was the effective thermal conductivity
λ was the latent heat of the vaporization of water

The unsteady-state momentum balance in turbulent conditions (i.e., the so-called Reynolds-averaged Navier-Stokes equations)—coupled to both the continuity equation and the transport equations for k and ω, the energy balance (accounting for both convective and conductive contributions), and the mass balance (accounting for both convective and diffusive contributions)—was used to model the transport phenomena in the drying air.

Momentum balance and continuity equation:

$$\frac{\partial \rho_a}{\partial t} + \underline{\nabla} \cdot \rho_a \underline{u} = 0 \tag{7.4}$$

$$\rho_a \frac{\partial \underline{u}}{\partial t} + \rho_a \underline{u} \cdot \underline{\nabla} \underline{u} + \underline{\nabla} \cdot \left(\overline{\rho_a \underline{u}' \otimes \underline{u}'} \right) = -\underline{\nabla} p + \underline{\nabla} \cdot \left[\eta_a \left(\underline{\nabla} \underline{u} + (\underline{\nabla} \underline{u})^T \right) \right] \tag{7.5}$$

$$\rho_a \frac{\partial k}{\partial t} + \rho_a \underline{u} \cdot \underline{\nabla} k = \underline{\nabla} \cdot \left[(\eta_a + \sigma_k \eta_t)(\underline{\nabla} k) \right] + \frac{\eta_t}{2} \left(\underline{\nabla} \underline{u} + (\underline{\nabla} \underline{u})^T \right)^2 - \beta_k \rho_a k \omega \tag{7.6}$$

$$\rho_a \frac{\partial \omega}{\partial t} + \rho_a \underline{u} \cdot \underline{\nabla} \omega = \underline{\nabla} \cdot \left[(\eta_a + \sigma_\omega \eta_t)(\underline{\nabla} \omega) \right] + \left(\frac{\alpha \omega}{2k} \right) \eta_t \left(\underline{\nabla} \underline{u} + (\underline{\nabla} \underline{u})^T \right)^2 - \beta \rho_a \omega^2 \tag{7.7}$$

where:
  p was the pressure within the drying chamber
  u was the averaged velocity field
  $\otimes$ was the outer vector product
  $\underline{u}'$ was the fluctuating part of the velocity field
  β_k, σ_k, σ_ω, α, and β were constants

Heat balance equation:

$$\rho_a C_{pa} \frac{\partial T_2}{\partial t} - \underline{\nabla} \cdot \left(k_a \underline{\nabla} T_2 \right) + \rho_a C_{pa} \underline{u} \cdot \underline{\nabla} T_2 = 0 \tag{7.8}$$

where:
  T_2 was the air temperature
  C_{pa} was its specific heat
  k_a was the air thermal conductivity (actually comprised by both a laminar
      and a turbulent contribution to energy transport by conduction)

Mass balance equation:

$$\frac{\partial C_2}{\partial t} + \underline{\nabla} \cdot \left(-D_a \underline{\nabla} C_2 \right) + \underline{u} \cdot \underline{\nabla} C_2 = 0 \tag{7.9}$$

where:
  C_2 was the water concentration, as vapor, in the air
  D_a was the diffusion coefficient of vapor in air

Also, in this case, both a laminar and a turbulent contribution were considered to express mass transfer due to a diffusion mechanism. It is worthwhile remarking that the physical and the transport properties of food and drying air were expressed in terms of the local values of both temperature and moisture content.

A set of initial conditions was necessary to perform the numerical simulations. As far as air was concerned, it was assumed that temperature in the drying chamber, T_a, had, initially, the value of 298 K and that the water concentration, C_a, was equal to 0.642 mol/m^3 (corresponding to a relative humidity, H, of 50%). It was also assumed that, before drying process occurred, air was stagnant (i.e., $\underline{u} = \underline{0}$) and that pressure in the drying chamber, p_0, was 1 atm. The initial values of food temperature, T_0, and of its moisture content, U_0, were set equal, respectively, to 283.15 K and to 4.4 kg H$_2$O/kg$_{ds}$ (on a dry basis). It is worthwhile remarking that the present transport model was flexible and allowed changing the values of T_a, H, air velocity, u_0, and T_0 and U_0, so as to simulate drying process behavior in a wide range of process conditions having a physical significance.

As far as boundary conditions were concerned, at the food-air interface, where no accumulation occurred, the continuity of heat and water fluxes was imposed. Moreover, a thermodynamic equilibrium at the air-food interface was formulated and expressed in terms of water activity, so to account also for the effects of physically bound water. The values of temperature, of vapor concentration, and of air velocity entering the oven were fixed before performing each simulation and represented three input parameters that could be deliberately changed in order to analyze system behavior over as wide a range as possible of process and fluid-dynamic conditions.

Moreover, it was assumed that at the dryer walls air temperature and its vapor concentration were equal to the corresponding values measured at the drier inlet section. These two conditions are valid under the assumption that temperature and concentration profiles were confined to two very thin regions, which developed close to the food-air interface. At the drier outlet, conduction and diffusion were neglected in favor of convection, which prevailed. Finally, the boundary conditions for momentum balance at the solid walls were expressed in terms of a logarithmic wall function.

It is worthwhile observing that the formulated theoretical model could be considered very general since it did not make use of any transport coefficient aimed at estimating the heat and mass fluxes at the food-air interfaces.

The described transport model, as represented by Equations 7.1 through 7.9 together with the corresponding boundary and initial conditions, was capable of describing the time evolution of the drying process but actually did not give any indication about the progress of product decontamination and the quality of dried vegetable. For this reason, two additional ANN models were developed and combined to the transport equations with the aim of relating the calculated local values of moisture content and temperature in food to microbial inactivation and color degradation, both depending on the set of operating conditions chosen to perform drying process. To develop the first ANN model (ANN1) aimed at estimating the progress of food decontamination, a set of experimental data reported in Valdramidis et al. (2006) was exploited to calculate microbial inactivation of *Listeria monocytogenes* as a function of food temperature, water activity, and treatment duration. A multilayer perceptron (MLP) feed-forward architecture, having a pyramidal structure and five neurons in the intermediate hidden layer, was eventually identified by MATLAB® Neural Network Toolbox (MathWorks), version 4.0.1. To predict the time evolution of color changes occurring during potato drying, the experimental data available in the papers published by Krokyda et al. were exploited (Krokida et al. 1998; Krokida et al. 2001; Krokida et al. 2003). A second multilayer feed-forward neural network (ANN2), having a pyramidal structure and seven neurons in the intermediate hidden layer, was developed to describe the color change kinetics in terms of the so-called Hunter parameters—redness (*a*), yellowness (*b*), and lightness (*L*). It was therefore possible to express Hunter parameters as a function of drying time and of the operating conditions—the dry bulb temperature, T_a, and the relative humidity, *H*, of drying air.

The theoretical model represented by the system of PDEs (Equations 7.1 through 7.9) was coupled to the neural models previously described. The resulting hybrid neural model, whose architecture is shown in Figure 7.2, allowed simulation of the potato drying process, thus identifying the corresponding microbial deactivation and color changes.

7.2.2 Results and Discussion

The proposed hybrid neural model allowed a determination, on the basis of a definite set of input parameters—dry-bulb temperature (T_a), relative humidity (*H*), and feed velocity (u_0) of drying air—regarding the actual time evolutions of both moisture content and temperature within the food.

Figures 7.3 and 7.4 show the time evolution of potato moisture content (on a dry basis) and of temperature, respectively. It could be observed that, as expected, external surfaces got dry more rapidly than inner regions. When, initially, food moisture content was high and, consequently, capillary diffusivity had large values, moisture content exhibited a slight difference

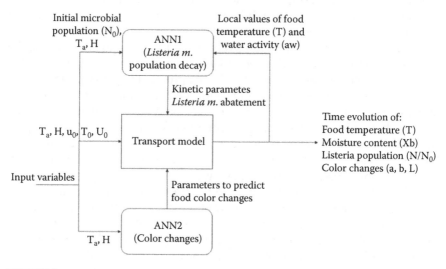

FIGURE 7.2
Hybrid neural model structure.

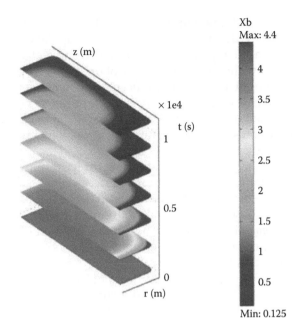

FIGURE 7.3
Time evolution of potato moisture content during drying ($T_a = 70°C$, $H = 30\%$, $u_0 = 2.2$ m/s).

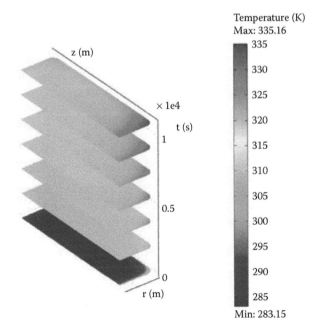

FIGURE 7.4
Time evolution of potato temperature during drying ($T_a = 70°C$, $H = 30\%$, $u_0 = 2.2$ m/s).

between the core and the outer surface. These differences tended to enlarge progressively due to both a decrease in capillary diffusivity and an increase of surface temperature that started rising well above wet-bulb temperature. The low thermal conductivity of dry regions on food surfaces, together with evaporation, led to a significant drop in temperature from the surface to the food core (Figure 7.4). During the falling-rate period, the evaporation rate prevailed over the capillary flux inside the food; the amount of vapor in internal pores was significant, and molecular diffusion of vapor took place to a significant extent.

The present HNM allowed also determining the local values of both water activity and temperature on food external surfaces, in a wide range of process and operating conditions that could be changed by varying the input parameters. In this way, it was possible to estimate the actual reduction of the *Listeria monocytogenes* population. In addition, it was possible to estimate the corresponding color changes as a function of process time and of the chosen values of T_a and H.

Figures 7.5 and 7.6 show, respectively, the calculated variations of redness, Δa, and yellowness, Δb, defined as the difference between the current value and the initial one. As drying proceeds, redness increases for any combination

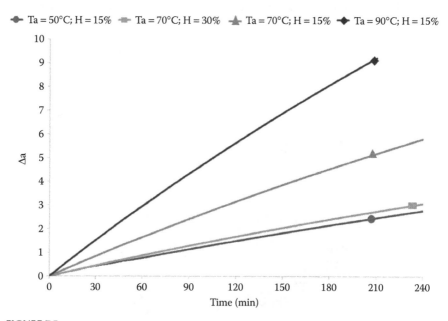

FIGURE 7.5
Potato color changes (redness) at different values of air dry-bulb temperature and relative humidity ($u_0 = 2.2$ m/s).

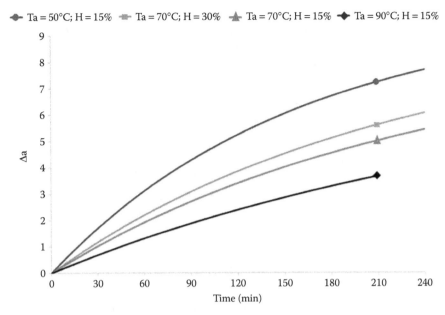

FIGURE 7.6
Potato color changes (yellowness) at different values of air dry-bulb temperature and relative humidity ($u_0 = 2.2$ m/s).

of T_a and H; in particular, a more pronounced rise of Δa is observed as drying temperature is augmented and relative humidity decreased. As far as yellowness is concerned, more evident color changes are observed as drying temperature decreases and, correspondingly, relative humidity increases. The effect of temperature on the variations of both redness and yellowness, however, is more intense than that of H.

Figure 7.7 shows, as a function of air relative humidity, the decrease of the *Listeria m.* population on a point laying on the food rear surface—where there exists a segregation region resulting in an inefficient air circulation. A significant variation of microbial population abatement could be observed as H was changed. In some of the tested cases, however, the microbial population was still large (about four orders of magnitude lower than the initial one), and this might result in an unsatisfactory decontamination of dried food.

Generally speaking, the exploitation of drastic operating conditions—high dry-bulb temperatures and low values of air relative humidity—has the following effects: (a) the drying rate is high, and the process goes to completion faster; (b) the reduction of the microbial population is more efficient, and food safety is improved; and (c) the organoleptic characteristics of food tend to

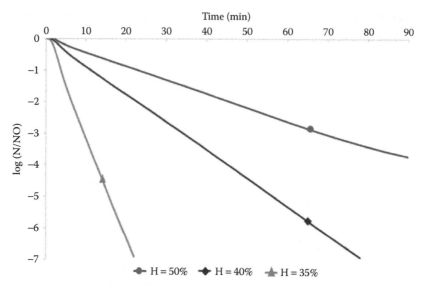

FIGURE 7.7
Time evolution of *Listeria* population during drying ($T_a = 70°C$, $u_0 = 2.2$ m/s).

deteriorate, thus determining a significant worsening of dried product quality. In contrast, the utilization of mild or very mild operating conditions—such as low dry-bulb temperatures and intermediate values of relative humidity— determines an opposite behavior: the organoleptic properties of dried foods are better, but a proper decontamination of the final product cannot be assured. One of the major practical advantages related to the proposed approach is that the developed hybrid neural model can be exploited to determine which combination of T_a and H may enhance, at the same time, the quality and the safety of dried vegetables. In this way, it is possible to reduce the number of pilot test runs, generally expensive and time consuming, that are necessary to achieve an experimentally based knowledge of the system under study and to determine the influence of operating variables on drying performance.

7.3 Case Study 2: Enzymatic Transesterification of Waste Olive Oil Glycerides for Biodiesel Production

7.3.1 Kinetic and Theoretical Model

The reaction pattern of biocatalytic transesterification of triolein (the triglyceride contained in olive oil) in the presence of ethanol was described in Calabrò et al. (2009) as a sequence of three reactions in series, which lead to the formation of one mole of ester for each step and to the obtainment of glycerol only at the third step, when monoglycerides are actually converted:

$$\text{Triolein + Ethanol} \leftrightarrow \text{Diolein + Ethyloleate}$$

$$\text{Diolein + Ethanol} \leftrightarrow \text{Monolein + Ethyloleate}$$

$$\text{Monolein + Ethanol} \leftrightarrow \text{Glycerol + Ethyloleate}$$

The proposed mechanism has been revised and simplified, considering tri-olein and ethanol as the substrates, with ethyloleate, glycerol, and the other glycerides (monolein and diolein) as the products. The complex kinetic mechanism eventually described by a Ping-Pong Bi-Bi mechanism with ethanol inhibition and the King-Altman kinetics method (King and Altman 1956), based on singling out geometrical rules that permits evaluating the concentrations of enzyme in all its complexes ($[E]$, $[e]$, $[ES]$, $[EP]$, etc.), was adopted.

By considering the actual rate of each elementary reaction, it was possible to formulate a simplified kinetic rate equation, expressed as the disappearance of triolein $[T]$, as follows (Calabrò et al. 2009):

$$v_T = -\frac{d[T]}{dt} = \frac{\alpha_1 \cdot [T] \cdot [Et] - \beta_1 \cdot [P] \cdot [EO]}{[T]^2 + \delta \cdot [T] + \varepsilon} \cdot [e_0] \tag{7.10}$$

where:
$\alpha_1, \beta_1, \delta, \varepsilon$ were kinetic constants
$[T]$ represented triolein concentration [mol/l]
$[Et]$ was ethanol concentration [mol/l]
$[P]$ was the overall concentration of glycerol, monolein, and diolein [mol/l]
$[EO]$ was ethyloleate concentration [mol/l]
$[e_0]$ was lipase concentration

A possible limitation of this kinetic analysis is actually represented by the methodology adopted to express, on the basis of stoichiometry and some semiempirical correlations, the concentrations of products and ethanol as a function of triolein actual concentration $[T]$ and of substrate initial concentrations $[T_0]$ and $[Et_0]$. In particular, it was assumed as follows (Calabrò et al. 2009):

$$[Et] = 2.25 \cdot ([T] - [T_0]) + [Et_0] \tag{7.11}$$

$$[EO] = -2.25 \cdot ([T] - [T_0]) \tag{7.12}$$

$$[P] = [T_0] - [T] \tag{7.13}$$

In the present contribution, a completely different methodology aimed at determining the actual relationship existing among the concentrations of products and ethanol and the variables $[T]$, $[T_0]$, and $[Et_0]$ is presented.

The already-determined linear relationship between triolein and ethylole-ate, as provided by Equations 7.11 through 7.13, in principle might not be accurately verified in some cases, especially when the concentrations of reactant(s) or of product(s) are low—at the beginning or at the end of the reaction. Product obtainment, in fact, exhibited in some circumstances an initial delay with reference to substrate consumption. In addition, a final decrease of product production rate as compared to substrate consumption rate was observed. An improper estimation of the actual substrate(s)-product(s) relationship, therefore, may lead to biased predictions of the biocatalytic reaction under study, especially if it is considered that the initial rate strongly affects the actual process dynamics, whereas the final values are critical when the reaction yield and the substrate conversion are to be calculated.

The methodology proposed in this chapter makes use of advanced computational models, based on artificial neural networks, properly integrated with the already-proposed kinetic mechanism so as to formulate an overall HNM, which is expected to provide more-reliable predictions of the actual time evolutions of substrate(s) and product(s) concentrations involved in the biocatalytic transesterification process. It is worthwhile observing that the Ping-Pong Bi-Bi mechanism was considered to be valid in the present case also. The simplified form expressed by Equation 7.10 will hereafter be exploited, since it permits reducing the model complexity while ensuring an acceptable description of the overall system kinetics.

7.3.1.1 Hybrid Neural Model Development

The inherent limitation of the kinetic model presented in Calabrò et al. (2009) was the identification of an empirical linear correlation, obtained by inter-polating the experimental data collected in specific operating conditions existing between the instantaneous product concentration $[EO]$ and the substrate instantaneous concentration $[T]$. With the aim of widening the validity domain of the kinetic equation (Equation 7.10), a hybrid neural model was developed. In particular, the relation existing between substrate and product concentrations was determined on the basis of an artificial neural network, properly trained with the experimental data obtained over a wide range of process and operating conditions. The theoretical part of the proposed HNM preserved kinetic Equation 7.10, but a neural network was exploited to predict ethyloleate concentration on the basis of substrate concentration values and reaction operating conditions.

In order to predict the neural network output variable, a set of significant input variables were considered as a result of a sensitivity analysis performed on the biocatalytic process under study:

- Enzyme/triolein initial mass ratio ($[e_0]/[T_0]$)
- Ethanol/triolein initial molar ratio ($[Et_0]/[T_0]$)

- Initial water content of the reaction mixture ($[W_0]$)
- Agitation rate of the reactor (rpm)
- Triolein/hexane initial ratio ($[T_0]/[Ex_0]$)
- Triolein consumption, expressed in terms of concentration ($[T_0]-[T(t)]$)

These variables, among all the others that could affect the reaction progress, exhibited the highest influence on product formation.

The input-output structure of the HNM is synthesized in Figure 7.8.

The realized HNM is characterized by a parallel structure, since a continuous recycle of a signal exists between the theoretical and the neural part of the model.

After specifying the input-output structure of the model, the neural network architecture was defined and the training procedure set up. Considering the available set of experimental data, the experimental runs were split in two groups, reserving two-thirds of reaction runs, corresponding to 88 experimental points, to the neural network training/validation phase. The remaining one-third of reaction runs, corresponding to 50 points, were used to test the predictions of the developed ANN in a set of conditions never exploited, either during learning or during validation. A multilayer perceptron (MLP) feed-forward architecture with a pyramidal structure was identified by MATLAB Neural Network Toolbox (MathWorks), version 4.0.1. For the training of each tested network, the Bayesian regularization technique was used (Demuth et al. 2000); it allows a significant improvement of ANN generalization. The neuron transfer function was a hyperbolic tangent for both input and hidden layers, whereas a linear transfer function was

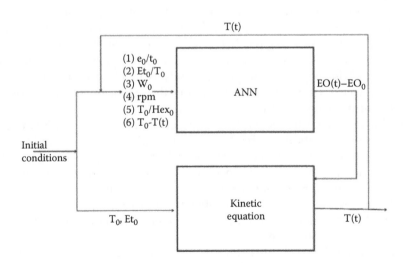

FIGURE 7.8
Input-output structure of the HNM.

chosen for the output layer. The choice of the *best* network architecture was achieved by a trial-and-error procedure, as suggested in the literature Curcio et al. (2005, 2010); the number of both the hidden layers and the neurons belonging to each layer were determined through iterative cycles. It was supposed that the convergence was achieved when, during the training/validation phase, the percentage average error between the predictions of neural model and the corresponding experimental points (Equation 7.14) reached a minimum, set equal to 5%.

$$\left(\varepsilon_A\% = \frac{|EO(t)_{\text{EXP}} - EO(t)_{\text{ANN}}|}{EO(t)_{\text{EXP}}} \cdot 100 \right) \tag{7.14}$$

On the basis of the foregoing iterative procedure, a neural model having two intermediate hidden layers with 10 and 3 neurons, respectively, was eventually developed.

7.3.2 Results and Discussion

The experimental data of produced ethyloleate ($[EO(t)]-[EO_0] = [EO]$) versus triolein consumption ($[T_0]-[T]$) were reported for each of the experiments carried out, in order to investigate the effect of operating conditions on neural network performance. The results provided by the present HNM have been compared (Figure 7.9) with those obtained in the previous study (Calabrò et al. 2009), where a linear relationship was found, as described by Equation 7.11 and 7.13.

It can be observed that the experimental points representing the ethyloleate versus triolein concentration do not exhibit a linear trend, as it was supposed in the previous paper (Calabrò et al. 2009). On the contrary, a pseudo-sigmoidal behavior can be identified. Consequently, the neural network model developed in the present study is much more reliable than a pure linear correlation in the prediction of the actual relationship existing between ethyloleate and triolein concentrations.

Having estimated the general relation between [EO] and [T], it was possible to test the reliability of the developed hybrid neural model and to compare its predictions with the experimental results. Figure 7.10 shows the concentration profiles of all the species, in a typical case. It can be observed that a significant agreement exists between the experimental points and the simulation results: the HNM is capable of predicting very accurately both the actual time evolutions and the plateau values for all the species involved in transesterification process.

The following two figures (Figures 7.11 and 7.12) show a comparison between the predictions provided by the present HNM and those obtained by the theoretical approach proposed in Calabrò et al. (2009), which was

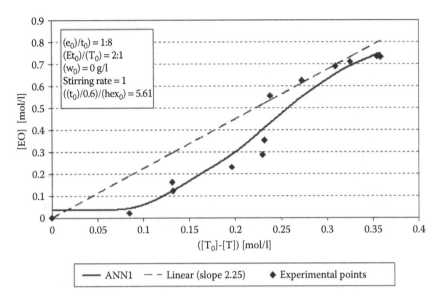

FIGURE 7.9
Comparison between experimental data, predictions provided by ANN, and linear empirical correlation ($[EO] = 2.25*([T_0]–[T]$) for the determination of the relationship between ethyloleate production and triolein consumption.

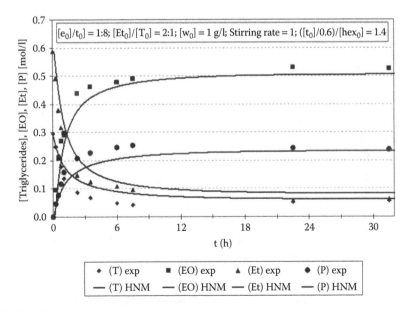

FIGURE 7.10
Comparison between HNM predictions and experimental data.

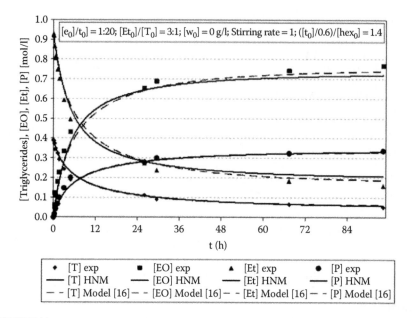

FIGURE 7.11

Comparison between HNM predictions, experimental data, and the predictions provided by the theoretical approach proposed in the previous study [16].

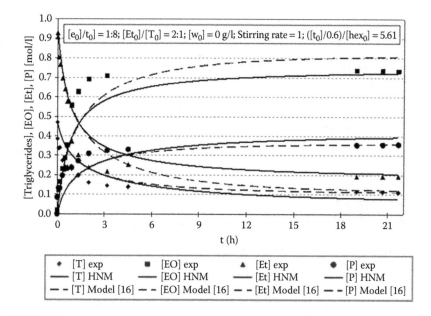

FIGURE 7.12

Comparison between HNM predictions, experimental data, and the predictions provided by the theoretical approach proposed in the previous study [16].

based on the assumption that a linear relationship existed between ethyloleate and triolein concentrations. It can be observed in both the considered cases that the predictions provided by the HNM are more accurate than those obtained in the previous case.

In particular, it can be noted that, at the beginning of the reaction run, both the HNM and the model described in Calabrò et al. (2009) were characterized by a certain level of inaccuracy and that both the models tended to underestimate the ethyloleate concentration. However, the approach proposed in Calabrò et al. (2009), being based on a best-fitting model, in the second part of the reaction run tended to overestimate the ethyloleate concentration, thus resulting in a worse prediction of ethyloleate plateau value. In contrast, the HNM, not being based on a simple interpolation of experimental data but rather on a more powerful computational approach, is capable of providing a more accurate prediction of the final values of ethyloleate concentration.

7.4 Conclusions

In this chapter, two different hybrid neural models were formulated with the aim of predicting the behavior of two different processes: the convective drying of vegetables and the enzymatic transesterification of glycerides contained in waste olive oil.

It was proven that the first HNM can be exploited to determine which set of operating conditions would enhance the quality and the safety of the final product, thus minimizing expensive and time-consuming pilot test runs. It is, in fact, possible to predict the spatial moisture profiles at all times, thus allowing the detection of the regions where either high values of moisture content or low values of temperature can promote microbial spoilage. The developed HNM might be used, in a future application, to perform a real optimization of a drying process so as to identify a trajectory of operating conditions that may allow achieving specific control objectives represented, for instance, by the determination of a trade-off condition between food quality and safety.

The second HNM demonstrated the ability to account for the influence of different input variables on the transesterification yield. For this reason, the developed HNM could successfully be used to realize a control system for the process, where operating conditions might be chosen as manipulating variables and yield or conversion might be used as output variables. This methodology offers a very interesting opportunity for the kinetic study of bioconversions, where the effects of some input variables are sometimes difficult to properly account for.

List of Symbols

a	*Redness*: Hunter parameter expressing color change	–
B	*Yellowness*: Hunter parameter expressing color change	–
C_2	Water concentration in air	mol/m³
C_{pa}	Air-specific heat	J/(kg K)
C_{ps}	Food-specific heat	J/(kg K)
C_v	Vapor concentration in food	mol/m³
C_w	Liquid water concentration in food	mol/m³
D_a	Diffusion coefficient of water vapor in air	m²/s
D_v	Effective diffusion coefficient of vapor in food	m²/s
D_w	Capillary diffusion coefficient of water in food	m²/s
$[e_0]$	Lipase concentration	[g/L]
$[EO]$	Ethyloleate concentration	mol/L
$[E_t]$	Ethanol concentration	mol/L
$[Ex_0]$	Hexane initial concentration	mol/L
H	Air relative humidity at the drier inlet	–
$\dot{I}$	Volumetric rate of evaporation	mol/(m³ s)
k	Turbulent kinetic energy	m²/s²
k_a	Air thermal conductivity	W/(m K)
k_{eff}	Effective thermal conductivity of food	W/(m K)
L	Lightness: Hunter parameter expressing color change	–
$\underline{n}$	Unity vector normal to the surface	–
N	Microbial population	cfu/ml
p	Pressure within the drying chamber	Pa
$[P]$	Overall concentration of glycerol, monolein, and diolein	mol/L
R	Gas constant	J/(mol K)
r	Radial coordinate	M
rpm	Agitation rate of the reactor	min⁻¹
T	Food temperature	K
t	Time	s
$[T]$	Triolein concentration	mol/L
T_2	Air temperature	K
T_a	Air temperature at the drier inlet	K
u	Averaged velocity field	m/s
u'	Fluctuating part of velocity field	m/s
$u0$	Air velocity in axial direction at the drier inlet	m/s
u_t	Friction velocity	m/s
$[W_0]$	Initial water content of the reaction mixture	mol/L
X_b	Moisture content on a dry basis	kg water/kg dry solid
z	Axial coordinate	m

Greek Symbols

α	Constant appearing in k-ω model	–
α_1	Kinetic constant in Equation 7.10	–
β	Constant appearing in k-ω model	–
β_1	Kinetic constant in Equation 7.10	
β_k	Constant appearing in k-ω model	–
δ	Kinetic constant in Equation 7.10	
Δ_a	Variation of potato redness with respect to the initial value	–
Δ_b	Variation of potato yellowness with respect to the initial value	–
ε	Kinetic constant in Equation 7.10	–
$\varepsilon_A\%$	Percentage average error	–
η_a	Air viscosity	Pa s
$\eta_t = \rho_a k/\omega$	Air turbulent viscosity	Pa s
κ	von Karman's constant	–
λ	Water latent heat of vaporization	J/mol
ρ_a	Air density	kg/m^3
ρ_s	Food density	kg/m^3
σ_k	Constant appearing in k-ω model	–
σ_ω	Constant appearing in k-ω model	–
ω	Dissipation per unit of turbulent kinetic energy	–

Subscripts and Superscripts

0	Initial condition ($t = 0$)
ANN	Calculated by neural network model
EXP	Measured
r	in radial direction
z	in axial direction

References

Agarwal, M. 1997. Combining neural and conventional paradigms for modelling, prediction and control. *International Journal of System Science* 28: 65.

Al-Mahasneh, M., M. Aljarrah, T. Rababah, and M. Alu'datt. 2016. Application of hybrid neural fuzzy system (ANFIS) in food processing and technology. *Food Engineering Reviews* 8: 351–366.

Andrić, P., A. S. Meyer, P. A. Jensen, and K. Dam-Johansen. 2010. Reactor design for minimizing product inhibition during enzymatic lignocellulose hydrolysis: I. Significance and mechanism of cellobiose and glucose inhibition on cellulolytic enzymes. *Biotechnology Advances* 28: 308.

Basile, A., S. Curcio, G. Bagnato, S. Liguori, S. M. Jokar, and A. Iulianelli. 2015. Water gas shift reaction in membrane reactors: Theoretical investigation by artificial neural networks model and experimental validation. *International Journal of Hydrogen Energy* 40: 5897–5906.

Calabrò, V., E. Ricca, M. G. De Paola, S. Curcio, and G. Iorio. 2009. Kinetics of enzymatic trans-esterification of glycerides for biodiesel production. *Bioprocess and Biosystems Engineering* 33: 701–710.

Chen, F. F., M. Breedon, P. White, C. Chu, et al. 2016. Correlation between molecular features and electrochemical properties using an artificial neural network. *Materials and Design* 112: 410–418.

Cubeddu, A., C. Rauha, and A. Delgado. 2014. Hybrid artificial neural network for prediction and control of process variables in food extrusion. *Innovative Food Science and Emerging Technologies* 21: 142–150.

Curcio, S. and M. Aversa. 2014. Influence of shrinkage on convective drying of fresh vegetables: A theoretical model. *Journal of Food Engineering* 123: 36–49.

Curcio, S., G. Scilingo, V. Calabrò, and G. Iorio. 2005. Ultrafiltration of BSA in pulsating conditions: An artificial neural networks approach. *Journal of Membrane Science* 246: 235–247.

Curcio, S., M. Aversa, and A. Saraceno. 2010. Advanced modeling of food convective drying: A comparative study among fundamental, artificial neural networks and hybrid approaches. In *Food Engineering*, Vol. 14, B. C. Siegler (Ed.). Hauppauge, NY: Nova Science Publishers, ISBN: 978-1-61728-913-2.

Curcio, S., V. Calabrò, and G. Iorio. 2009. Reduction and control of flux decline in cross-flow membrane processes modeled by artificial neural networks and hybrid systems. *Desalination* 236: 234.

De Paola, M. G., E. Ricca, V. Calabrò, S. Curcio, and G. Iorio. 2009. Factor analysis of transesterification reaction of waste oil for biodiesel production. *Bioresource Technology* 100: 5126–5131.

Demuth, H. and M. Beale. 2000. *Neural Network Toolbox User's Guide*. Natick, MA: The MathWorks.

Estiati, I., F. B. Freire, J. T. Freire, R. Aguado, and M. Olazar. 2016. Fitting performance of artificial neural networks and empirical correlations to estimate higher heating values of biomass. *Fuel* 180: 377–383.

Grahovac, J., A. Jokić, J. Dodić, D. Vučurović, and S. Dodić. 2016. Modelling and prediction of bioethanol production from intermediates and byproduct of sugar beet processing using neural networks. *Renewable Energy* 85: 953–958.

King, E. L. and C. Altman. 1956. A schematic method of deriving the rate laws for enzyme-catalyzed reactions. *Journal of Physical Chemistry* 60: 1375–1378.

Krokida, M. K., V. T. Karathanos, Z. B. Maroulis, and D. Marinos-Kouri. 2003. Drying kinetics of some vegetables. *Journal of Food Engineering* 59: 391–403.

Krokida, M. K., Z. B. Maroulis, and G. D. Saravacos. 2001. The effect of the method of drying on the colour of dehydrated products. *International Journal of Food Science and Technology* 36: 53–59.

Krokida, M. K., E. Tsami, and Z. B. Maroulis. 1998. Kinetics on colour changes during drying of some fruits and vegetables. *Drying Technology* 16: 667–685.

Neethu, K. C., A. K. Sharma, H. A. Pushpadass, F. M. Eljeeva Emerald, and M. Manjunatha. 2016. Prediction of convective heat transfer coefficient during deep-fat frying of pantoa using neurocomputing approaches. *Innovative Food Science and Emerging Technologies* 34: 275–284.

Reilly, D. L. and L. N. Cooper. 1990. An overview of neural networks: Early models to real world systems. In S. F. Zornetzer, J. L. Davis, and C. Lau (Eds.), *An Introduction to Neural and Electronic Networks*, New York: Academic Press.

Saraceno, A., S. Curcio, V. Calabrò, and G. Iorio. 2010. A hybrid neural approach to model batch fermentation of "ricotta cheese whey" to ethanol. *Computers and Chemical Engineering* 34: 1590–1596.

Saraceno, A., S. Sansonetti, V. Calabrò, G. Iorio, and S. Curcio. 2011. A comparison between different modeling techniques for the production of bio-ethanol from dairy industry wastes. *Chemical and Biochemical Engineering Quarterly* 25: 461–469.

Simutis, R., M. Dors, and A. Lubbert. 1995. Bioprocess optimization and control: Application of hybrid modelling. *Journal of Biotechnology* 42: 285.

Valdramidis, V. P., A. H. Geeraerd, J. E. Gaze, A. Kondjoyan, A. R. Boyd, H. L. Shaw, and J. F. Van Impe. 2006. Quantitative description of *Listeria monocytogenes* inactivation kinetics with temperature and water activity as the influencing factors: Model prediction and methodological validation of dynamic data. *Journal of Food Engineering* 76: 79–88.

Zhang, G., B. E. Patuwo, and M. J. Hu. 1998. Forecasting with artificial neural network: The state of art. *International Journal of Forecasting* 14: 35.

8

Hybrid Modeling of Pharmaceutical Processes and Process Analytical Technologies

Jarka Glassey

CONTENTS

8.1 Quality by Design and Process Analytical Technologies

The significance of the pharmaceutical and biopharmaceutical sector within the worldwide socioeconomic context is undoubtedly widely accepted due to its role in maintaining and improving the health of the population. The pharmaceutical industry was valued at $778 billion from the total worldwide sales of prescription and over-the-counter drugs in 2016, with 25% of this revenue generated by biological/biotechnological products (Pharma 2016). The consensus forecast of worldwide prescription drug sales is set to exceed $1 trillion, reaching $1.12 trillion by 2022 and showing annual compound growth of 6.3% from 2016 to 2022 (Pharma 2016). Based on historical data, a shift toward biologics seems imminent, owing to increasing profits and lower attrition rates when compared to small-molecule drugs. Biological drugs comprised

70% of the top-10-selling products of the world in 2014, and the percentage of sales of biotechnology products within the top 100 was 44%. The predictions for the future are that biologics will contribute 50% of the top 100 product sales by 2022 (Pharma 2016). Twenty new biologicals were approved by the U.S. Food and Drug Administration (FDA) in 2014 compared to the 11 that were approved in 2009. New drug approvals in 2015 included a record 56 new molecular entities (NMEs) (Pharma 2016). Monoclonal antibodies have higher approval rates, of 26%, in the biopharmaceutical sector than those of conventional small-molecule drugs (10%) (Hay et al. 2014).

It is recognized that the pharmaceutical and biopharmaceutical industries have lagged behind other industrial sectors in terms of innovative manufacturing engineering approaches (Rantanen and Khinast 2015). However, the introduction of quality by design (QbD) and process analytical technologies (PAT) initiatives, championed by regulatory bodies such as the FDA (2004), provided a strong impetus for more risk-based and knowledge-driven process development and production in these sectors.

QbD can be seen as a comprehensive approach to product development that includes designing and developing processes and identifying critical quality attributes, critical process parameters, and sources of variability (see Figure 8.1 for a graphical illustration of a QbD workflow). This approach aims to improve the understanding of how critical quality attributes (CQAs) are influenced by critical process parameters (CPPs) and the interactions between them (Kelley 2009). As part of the FDA's initiative "Pharmaceutical cGMPs for the Twenty-First Century: A Risk-Based Approach," it also produced a further document, "Guidance for Industry: PAT—A Framework for Innovative Pharmaceutical Development, Manufacturing, and Quality Assurance" (FDA 2004), in order to promote the introduction of new technologies to improve the efficiency and effectiveness

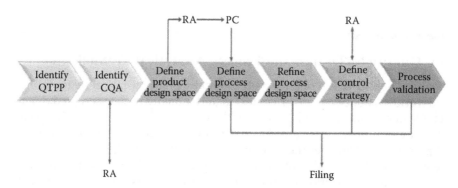

FIGURE 8.1
A schematic for quality by design in biopharmaceutical industry. RA: risk assessment, PC: process characterization. (Adapted from Rathore, A. S. and H. Winkle, *Nat Biotechnol*, 27 (1), 26–34, 2009.)

of manufacturing process design, quality control, and assurance. For example, Wu et al. (2015) argue that the FDA has been working in three specific areas to support new approaches for the improvement of product quality and manufacturing:

1. Continuous manufacturing with constant flow of materials in and out of a process
2. The use of PAT to monitor and control processes
3. Development of new statistical approaches to detect changes in process and product quality (Wu et al. 2015)

Despite marked advances in various aspects of biopharmaceutical process development and operation described in subsequent sections of this chapter, QbD is also reportedly seen as "costly, laborious and without the promised benefits of being able to freely act in the design space and do real-time release without regulatory oversight" (Herwig 2016). Arguably, the implementation of hybrid modeling techniques (alongside the actions proposed by Herwig 2016) as described in the case study detailed in Section 8.2 can contribute to addressing this negative perception.

8.1.1 Methodologies for Improved Modeling, Monitoring, and Optimization of Biopharmaceutical Processes

The importance of modeling within bioprocess development appears to have increased at the point of the QbD and PAT introduction. The biopharma industry has shown a greater interest in modeling (Velayudhan and Menon 2007), as the use of models has been shown to enhance the understanding of complex bioprocesses within pharmaceutical manufacturing, leading to long-term benefits with regard to more-rapid design and development of processes. In addition, accurate process models have been shown to be of value in on-line monitoring and control.

It is not possible to comprehensively review all of the different modeling approaches applied to all aspects of pharmaceutical and biopharmaceutical manufacturing sectors within the limits of the scope of this chapter, and thus the remainder of this chapter will focus more on the biopharmaceutical sector, given the increasing importance of biologics highlighted in Section 8.1. Readers are referred to specialist literature for details of other aspects of modeling (e.g., Wassgren et al. 2011; Benyahia et al. 2012; Dubey et al. 2012; Lakerveld et al. 2013; Rogers et al. 2013; Sen et al. 2014). For example, Rogers and Ierapetritrou (2015) provide a useful overview of advances in pharmaceutical process modeling, with emphasis on continuous manufacturing of oral-solid dosage forms as another significant segment of the pharmaceutical market in the United States.

8.1.1.1 First-Principles Modeling

The benefits of first-principles models have been argued clearly in this book, and Almquist et al. (2014) reinforce this in the area of industrial biotechnology. They argue that to account for the time dependence of the biological processes, the model relies on accurate predictions of the rates of production and consumption of each component. This is further supported by the models suggested by Kontoravdi et al. (2007), Naderi et al. (2011), and Xing et al. (2010), which involve developing mathematical representations for the metabolism modeling of CHO and various other cell cultures with different methods of rate prediction. Jang and Barford (2000) described an unstructured model for hybridoma cells, taking into account each of the main metabolites individually and also segregating viable and nonviable cells. This resulted in a more accurate prediction of the antibody production rate. A similar approach was adopted by Xing et al. (2010) and Naderi et al. (2011), with further developments being suggested by Kontoravdi et al. (2007) describing an unstructured model containing kinetic expressions for all metabolites (see Section 8.2). This increasing granularity and detail in the models is enabled by improved analytical measurements and increasing understanding of cell metabolism. Thanks to the *omics* data on the genome and the metabolome, these types of models allow the calculation of fluxes and have been used for media design and cell-line engineering and cultivation evaluation (Nolan and Lee 2011; Dietmair et al. 2012; Selvarasu et al. 2012).

However, the shortcomings of the first-principles models highlighted earlier are evident in this area of modeling. For example, Gadgil (2015) states that current mathematical models for animal cell growth and metabolism cannot simulate changes to the culture pH, with this being true for all process operating conditions. It can easily be seen from literature that changing culture conditions can greatly affect the performance of a cell culture (Ozturk et al. 1992; Yoon et al. 2005; Trummer et al. 2006; Ahn et al. 2008; Hwang et al. 2011; Li et al. 2012). In such situations, data-based modeling approaches may be able to capture the underlying dynamics, as discussed in the next section.

8.1.1.2 Data-Based Modeling of Bioprocesses

Rapid developments in analytical methodologies over recent years, partly resulting from the PAT initiative, led to a wealth of process data being recorded during bioprocess operations. As a result, increasingly, data-based modeling was more frequently applied as a method of choice in bioprocess modeling. For example, Glassey (2016) reviews various analytical methods and data-based modeling techniques for both feature extraction and regression; hence only a brief overview of relevant literature sources is reported here.

The application of QbD and PAT principles to laboratory-scale cultivations of bacterial expression systems frequently used in biopharmaceutical production has been reported extensively (e.g., in Carrondo et al. 2012, Mercier et al. 2013, and literature sources quoted in Chapter 5). In addition (Teixeira et al. 2009; Glassey et al. 2011; Carrondo et al. 2012; Mercier et al. 2013; Pais et al. 2014), and others have shown that multivariate data analysis (MVDA) can be used to extract important process information from a cell-culture data set.

However, arguably the most significant benefits arise from combining the benefits of both first-principles and data-based modeling approaches in a hybrid model as described in the following.

8.1.1.3 Hybrid Modeling

The drive to increase process understanding (not just since the introduction of QbD and PAT initiatives) drives the research in biopharmaceutical modeling inexorably towards hybrid modeling. For example, Gerdtzen (2012) argues that the answer to addressing the limitations of first-principles models in their failure to accurately capture responses to changes in operating conditions (see Section 8.1.1.1) is to develop models capable of capitalizing on data analysis in order to become adaptive to changes both in metabolism and operating conditions—a principal strength of hybrid models in this area.

Over the years, a number of hybrid models were developed for a range of microbial processes, some of which are reviewed in Chapter 5. More recently, research on important bioproducts such as amino acids (Brunet et al. 2012) or enzymes and other recombinant proteins produced by *E. coli* (Geethalakshmi et al. 2012; von Stosch et al. 2016) has reinforced this message. For example, Geethalakshmi et al. (2012) used a hybrid neural network of an unstructured first-principles model with a neural network to describe the feeding regime for the post-induction phase of a lab-scale fed-batch recombinant streptokinase production in order to increase enzyme activity. They achieved this through a segregated unstructured model describing the growth of both plasmid-bearing and plasmid-free *E. coli* cells, specifying separate growth rates for each. They subsequently trained a feed-forward neural network using biomass (both plasmid-bearing and plasmid-free), product and substrate concentrations as inputs to predict the rate parameters of the unstructured first-principle model, which then predicted the step-ahead biomass, product, and substrate values. The authors concluded that with a limited number of experiments, the hybrid neural network model was able to predict the state variables and identify the batch with the highest volumetric activity of the enzyme. Although the feeding regime was not actually predicted in the research described in this report, such extension of the model can be envisaged, given appropriate model input-output and parametrization.

In cell-culture hybrid modeling there were a number of research reports dealing with various aspects of cell-culture processes. For example, Popp et al. (2016) discuss the use of a hybrid model based on multivariate data analysis and a genome-based CHO network model comprising five compartments—cytosol, mitochondria, endoplasmic reticulum, Golgi apparatus, and the bioreactor—to identify metabolic signatures of high-producing CHO clones (Popp et al. 2016). Ten recombinant CHO-K1 clones were cultivated and characterized in detail by a set of indicators followed throughout the cultivation. The temporal metabolic changes and resulting signatures enabled high producers to be identified, which would significantly contribute to process optimization. As the authors envisage, in combination with high-throughput analytics, this modeling approach will be able to automatically rank the clones on the basis of indicators other than just the traditionally used titer. This could then be extended to product quality–related indicators such as product aggregation or glycosylation, although they acknowledge that this would require a more elaborate network model and refined mechanistic analysis of additional data sets in order to characterize the relationship between the cellular physiology and the product quality attributes in question. Such models can then be used, according to Popp et al. (2016), for on-line process monitoring and product quality control during cultivation, as discussed by Tharmalingam et al. (2015).

8.2 Case Study

The benefits of a semiparametric hybrid modeling approach within the biopharmaceutical sector are demonstrated here using the example of a monoclonal antibody (mAb) production with a cell-culture cultivation process, given the rising significance of the mAbs in the biopharmaceutical market (see Section 8.1). The challenges associated with mAb production also provide a significant challenge for the benefits of the various modeling approaches to be explored.

The structure of mAbs is studied extensively, and Figure 8.2 shows the basic structure containing four polypeptide chains—two light and two heavy chains. Figure 8.2 highlights the variable amino acid sequence in the Fab region of antigen-specific binding (light chain in blue and heavy chain in beige), the flexible hinge region, and the Fc region (constant for all antibodies of the same class). This part of the molecule facilitates interaction with the immune system once the antigen has bound to the Fab region.

Additional features are critical to the functionality of mAbs—particularly the glycosylation of the Fc region, which is essential for effector functions of

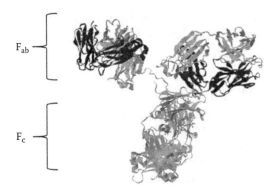

FIGURE 8.2
Structure of a neutralizing human IGG against HIV-1, showing the light chain in dark black color and heavy chain in gray color. The glycosylation is indicated as stick and ball. Structure available from PDB, http://www.rcsb.org/pdb/explore/explore.do?structureId=1HZH) under ID "1HZH." (Courtesy of Ollmann Saphire, E. et al., *Science*, 293 (5532), 1155–1159, 2001.)

the antibody (shown in Figure 8.2 as a stick and ball). Kizhedath et al. (2016) discuss in detail the mechanism and the consequences of any changes in the glycosylation profile on the biological function and/or pharmacokinetic and pharmacodynamic properties of the mAbs.

Given the importance of both the production process and CQA control during the process (in this case, control of the glycosylation profile), the case study presented in this chapter describes various modeling approaches aimed at increasing the yields of mAb and providing real-time information on glycosylation-resulting profiles in a hybridoma cell-culture cultivation process.

The cultivation experimentation was conducted by Ivarsson et al. (2014), and more-extensive details on the modeling used in this case are described in Green (2015). Briefly, the experiments were conducted using a hybridoma cell line (ATCC CRL-1606) in controlled parallel 1L bioreactors (DasGip) in batch mode. The culture conditions are reported in Ivarsson et al. (2014) and include the exploration of the effect of a parameter shift during the process. The temperature was controlled at 37°C, dissolved oxygen (DO) was set to 50% air saturation and controlled by a constant gas inlet flow rate of 0.05 vvm (volume of air per unit of medium per minute), pH was controlled at 7.2 by CO_2 sparging, stirrer speed was set to 150 rpm (revolutions per minute), and osmolality was 320 mOsm/kg. A parameter shift of one of the selected process variables was performed in the early exponential growth phase as described by Ivarsson et al. (2014). Viable cell concentration, glucose, lactate, and ammonia concentrations, and amino acid and mAb concentrations were measured twice a day as off-line data.

The glycosylation profile was recorded at the end of the cultivation, and all the sampling procedures and analytical methods are detailed in Ivarsson et al. (2014).

8.2.1 Feature Extraction

The initial multivariate data analysis of the recorded process data demonstrated that feature extraction methods such as parallel factor analysis (PARAFAC), principal component analysis (PCA), and multiway PCA (MPCA) were able to capture the variability contained within the process data, including the shift in the process parameters. For example, the biplots of the yield and the operating parameters (sparging, pH, DO, and osmolality) shown in Figure 8.3 and that of the glycosylation profile with the same operating parameters (Figure 8.4) indicate the correlation between these important performance indicators and the individual operating parameters (Green and Glassey 2015).

Although detailed analysis of the results is provided by Green (2015) and Green and Glassey (2015), it is worth highlighting the positive correlation

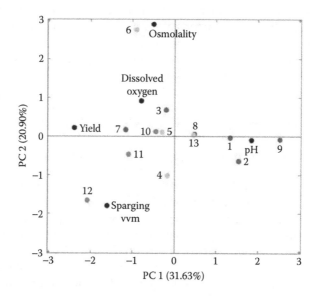

FIGURE 8.3
Biplot showing PC1 and PC2 for the PCA analysis of the off-line data for operating parameters and yield. The scores for individual batches are indicated by the numbers next to each symbol: dissolved oxygen (1, 2, and 3), osmolality (4, 5, 6, and 8), pH (7 and 9), and sparger (10, 11, 12, and 13). The variable loadings are shown in black and labeled with the operating parameter they relate to.

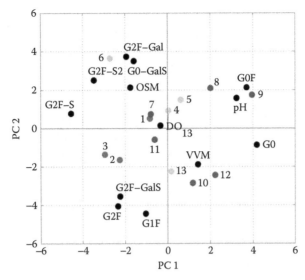

FIGURE 8.4

Biplot showing PC1 and PC2 for the PCA analysis of the off-line data for operating parameters and final product glycosylation profile. The scores for individual batches are indicated by the numbers next to each symbol: dissolved oxygen (1, 2, and 3), osmolality (4, 5, 6, and 13), pH (7, 8, and 9), and sparger (10, 11, and 12). The loadings are shown in black for glycans (represented by established nomenclature as detailed in A. Green and J. Glassey 2015, *Journal of Chemical Technology and Biotechnology* 90, 303–313) and the operating parameters labeled.

between the dissolved oxygen levels, osmolality, and yield and the negative correlation between pH and yield shown in Figure 8.3, by the positioning of the yield loadings in the same quadrant as the former two operating parameters and in the opposite quadrant to the pH loadings. Similarly, Figure 8.4 indicates some interesting correlations between the shifts in the operating parameters and the resulting glycan composition that correlate well with observations in the literature—for example, on the effect of the cultivation medium osmolality on the fucose content of the monoclonal antibodies (Konno et al. 2012; Ivarsson et al. 2014). The impact of the cultivation medium composition upon the glycosylation profile of the mAb is further demonstrated in the Figure 8.5 biplot.

The close clustering of the individual amino acid loadings and their relatively low values indicate a low correlation of individual amino acid concentrations and the glycosylation profile, although tryptophan (TRP) and glutamine (GLN) loadings indicate a greater correlation, in line with observations reported by Taschwer et al. (2012).

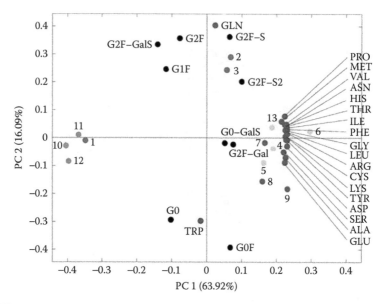

FIGURE 8.5

Biplot showing PC1 and PC2 for the PCA analysis of the off-line data for the amino acid concentrations and final product glycosylation profile. The scores for individual batches are indicated by the numbers next to each symbol: dissolved oxygen (1, 2, and 3), osmolality (4, 5, 6, and 13), pH (7, 8, and 9), and sparger (10, 11, and 12). The loadings are shown in black for glycans (represented by established nomenclature as detailed in A. Green and J. Glassey 2015, *Journal of Chemical Technology and Biotechnology* 90, 303–313) and in dark grey (with labels next to each symbol) for amino acids.

8.2.2 Prediction of Process Outcomes

Although the MVDA analyses described in Section 8.2.1 may contribute to a better understanding of the studied processes, from a pragmatic point of view, the ability to predict important process outcomes such as the concentrations of the viable cells and the product or the product quality in terms of its glycosylation profile are arguably equally (if not more) important from the manufacturing process operation point of view.

Thus regression MVDA techniques were subsequently used in this case study to predict the viable cell count, the product yield, and the glycosylation pattern at the end of the cultivation. The predictions of developed partial least squares (PLS) models are briefly reported here to enable a comparison of their performance with semiparametric hybrid models reported in Section 8.2.3.

Three types of models were developed for each of the process output variables considered, depending on the level of information used for model development:

a. Models based on set points of the operating parameters investigated (sparging, pH, DO, and osmolality)

b. Models based on off-line glucose and lactose measurement throughout the cultivation

c. Models based on on-line data (DO, flow rates of O_2 and CO_2, pH, and base addition)

The performance of the models for viable cell prediction on two validation batches is indicated in Figure 8.6, and the performance of each of the models for each of the output variables is summarized next in terms of the root

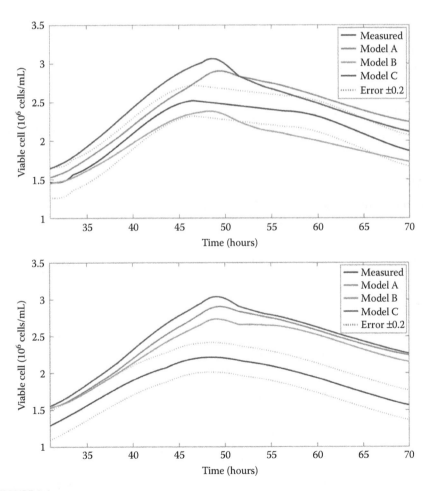

FIGURE 8.6
Prediction of viable cell count for two validation batches using PLS models described in Section 8.2.2 detailing the structure of the various PLS models.

mean square error (Equation 8.1) and Akaike's Information Criterion (AIC, Equation 8.2) as measures of the model prediction accuracy and complexity.

$$RMSE = \sqrt{\frac{\sum_{i=1}^{n}\left(y_i - \hat{y}_i\right)^2}{n}} \tag{8.1}$$

where:
y_i represents the measured output
$\hat{y}_i$ is the estimated value of the output variable over all n measurements

$$AIC = 2k + n\log\left(\frac{RSS}{n}\right) \tag{8.2}$$

where:
n is the number of observations
RSS is the Residual Sum of Squares
k is the number of model parameters

As Table 8.1 indicates, the performance of some of the PLS models in each output parameter estimation was relatively low in terms of RMSE, but

TABLE 8.1

Summary Information on Three Types of PLS Models Developed for the Prediction of Viable Cell Count, Product Yield, and the Glycosylation Profile

Parameter	Model	Input Data	RMSE Train.	RMSE Valid.	AIC Train.	AIC Valid.
Viable cell count	A	Operating parameters	0.28	0.43	11.57	4.73
	B	Glucose and lactose	0.24	0.21	14.36	12.12
	C	On-line data	0.23	0.36	15.20	17.04
Product yield	A	Operating parameters	7.29	6.12	27.30	29.46
	B	Glucose and lactose	6.84	9.88	29.77	35.76
	C	On-line data	6.57	9.57	28.42	35.59
Glycosylation	A	Operating parameters	0.83	1.37	0.87	1.10
	B	Glucose and lactose	0.63	1.40	1.90	3.51
	C	On-line data	0.62	1.31	0.97	3.50

Note: RMSE and AIC values (defined by Equations 8.1 and 8.2, respectively) are used as model quality measures.

Figure 8.6 indicates that despite relatively low RMSE error values, the estimations can be significantly different to the measured values, with the estimates of all three types of models exceeding the average experimental error range associated with this measurement. Similar performance was observed for the product yield and quality prediction, suggesting that a hybrid modeling approach may provide a more accurate prediction.

8.2.3 Hybrid Model Predictions of Process Outcomes

The results of PLS models developed to predict the viable cell count and mAb yield and glycosylation, presented in Section 8.2.2, indicated some room for improvement in the predictions (see Figure 8.6 for viable cell predictions). The first-principle models used in this research (Naderi et al. 2011; Kontoravdi et al. 2007) were also unable to provide more-accurate predictions, as Table 8.2 indicates, for the viable count and product yield predictions.

The Naderi et al. (2011) model was initially used as a simpler approach, relying on the basic metabolites (glucose, lactate, glutamine, ammonia, glutamate, asparagine, aspartate, and alanine). The more comprehensive model developed by Kontoravdi et al. (2007) includes the mass balances for all 20 amino acids involved in the metabolism and is summarized in Equations 8.3 through 8.31:

$$\frac{dV}{dt} = F_{in} + F_{glc} - F_{out} \tag{8.3}$$

$$\frac{d(VX_V)}{dt} = \mu VX_V - \mu_d VX_V - F_{out}X_V \tag{8.4}$$

TABLE 8.2

Model Predictions of Viable Cell Count and Product Yield Using First-Principle Models Compared to Best-Performing PLS Models Described in Section 8.2.2

Parameter	Model	Model Description	RMSE Train.	Valid.
Viable cell count	1	Naderi et al. (2011)	0.49	0.18
	2	Kontoravdi et al. (2007)	0.58	0.22
	PLS	Model B, Table 8.1	0.24	0.21
Product yield	1	Naderi et al. (2011)	17.56	14.04
	2	Kontoravdi et al. (2007)	14.43	10.13
	PLS	Model A, Table 8.1	7.29	6.12

$$\frac{d(VX_d)}{dt} = \mu_d V X_V - K_{lysis} V X_d - F_{out} X_d \tag{8.5}$$

$$X_t = X_V + X_d \tag{8.6}$$

The specific growth rate μ is also defined in this model in the context of the various amino acids involved in the growth of the CHO cell line as follows:

$$\mu = \mu_{min} + (\mu_{max} - \mu_{min}) \left[\frac{[GLC][GLN]}{(K_{glc} + [GLC])(K_{gln} + [GLN])} \right.$$

$$+ \frac{[ARG][VAL][LYS][THR]}{(K_{arg} + [ARG])(K_{val} + [VAL])(K_{lys} + [LYS])(K_{thr} + [THR])} \tag{8.7}$$

$$\left. \times \frac{[HIS][SER][ILE][PHE][LEU]}{(K_{his} + [HIS])(K_{ser} + [SER])(K_{ile} + [ILE])(K_{phe} + [PHE])(K_{leu} + [LEU])} \right]$$

For each of the amino acids (and the remaining metabolites), the mass balance is expressed as indicated for alanine ([ALA]) in Equation 8.8.

$$\frac{d(V[ALA])}{dt} = Q_{ala} V X_V + F_{in}[ALA]_{in} - F_{out}[ALA] \tag{8.8}$$

$$Q_{ala} = -\frac{\mu}{Y_{x,ala}} + Y_{ala,x} \tag{8.9}$$

However, the specific production/consumption rates for each of the amino acids varied based on the experimentally developed network of reactions reported by the authors and shown in Equations 8.10 through 8.28 for the remaining amino acids.

$$Q_{arg} = -\frac{\mu}{Y_{x,arg}} + Y_{arg,glu}Q_{glu} + Y_{arg,pro}Q_{pro} - Y_{arg,asp}Q_{asp} \tag{8.10}$$

$$Q_{asn} = -\frac{\mu}{Y_{x,asn}} - Y_{asn,asp}Q_{asp} \tag{8.11}$$

$$Q_{asp} = -\frac{\mu}{Y_{x,asp}} - Y_{asp,arg}Q_{arg} + Y_{asp,x} \tag{8.12}$$

$$Q_{cys} = -\frac{\mu}{Y_{x,cys}} - Y_{cys,ser}Q_{ser} \tag{8.13}$$

$$Q_{glu} = -\frac{\mu}{Y_{x,glu}} + Y_{glu,pro}Q_{pro} - Y_{gly,his}Q_{his} - Y_{glu,gln}Q_{gln} - Y_{glu,arg}Q_{arg} + Y_{glu,x} \tag{8.14}$$

$$Q_{gln} = -\frac{\mu}{Y_{x,gln}} - M_{gln} + Y_{gln,glu}Q_{glu} \tag{8.15}$$

where:

$$M_{gln} = \frac{\alpha_1[GLN]}{\alpha_2 + [GLN]} \tag{8.16}$$

$$Q_{gly} = -\frac{\mu}{Y_{x,gly}} - Y_{gly,ser}Q_{ser} \tag{8.17}$$

$$Q_{his} = -\frac{\mu}{Y_{x,his}} \tag{8.18}$$

$$Q_{ile} = -\frac{\mu}{Y_{x,ile}} \tag{8.19}$$

$$Q_{leu} = -\frac{\mu}{Y_{x,leu}} \tag{8.20}$$

$$Q_{lys} = -\frac{\mu}{Y_{x,lys}} + Y_{lys,x} \tag{8.21}$$

$$Q_{met} = -\frac{\mu}{Y_{x,met}} \tag{8.22}$$

$$Q_{phe} = -\frac{\mu}{Y_{x,phe}} \tag{8.23}$$

$$Q_{pro} = -\frac{\mu}{Y_{x,pro}} + Y_{pro,glu}Q_{glu} - Y_{pro,arg}Q_{arg} \qquad (8.24)$$

$$Q_{ser} = -\frac{\mu}{Y_{x,ser}} + Y_{ser,gly}Q_{gly} \qquad (8.25)$$

$$Q_{thr} = -\frac{\mu}{Y_{x,thr}} \qquad (8.26)$$

$$Q_{tyr} = -\frac{\mu}{Y_{x,tyr}} - Y_{tyr,phe}Q_{phe} \qquad (8.27)$$

$$Q_{val} = -\frac{\mu}{Y_{x,val}} \qquad (8.28)$$

Finally, for the other growth-related nutrients and metabolites (glucose, lactose, and ammonia), Equations 8.29 through 8.31 describe the specific rates of consumption/production as follows:

$$Q_{glc} = -\frac{\mu}{Y_{x,glc}} - M_{glc} \qquad (8.29)$$

$$Q_{lac} = -Y_{lac,glc}Q_{glc} \qquad (8.30)$$

$$Q_{amm} = -Y_{amm,gln}Q_{gln} \qquad (8.31)$$

Given the inaccuracies of both first-principles and MVDA regression models reported earlier, a hybrid approach indicated in Figure 8.7 was used to attempt to improve the model performance.

As Figure 8.7 indicates, PLS models were used with the initial concentrations of metabolites, cell count, and product yield together with the set points of the operating parameters to predict the rates of production and consumption of metabolites and the specific growth-rate values used in the first-principle models. Various combinations of MVDA and first-principle models were tested, as indicated in Table 8.3 and shown in Figure 8.8 for the same batches for viable cell-count prediction.

It is clear from this discussion (and Table 8.2 for comparison) that the viable cell-count predictions are significantly improved in some cases compared to the first-principles or PLS models used individually. Similar observations were made in the case of the product yield predictions, where a significant improvement compared to the first-principles models in particular was observed (data not shown).

FIGURE 8.7
Hybrid model structure indicating the modeling sequence. PLS models used $[C_0]$—the initial concentrations of metabolites, cell count, and product yield, together with operating parameter (OP) set points to predict the rates of production and consumption of metabolites and the cell growth rate. These are subsequently used in the mass balance calculations given by Kontoravdi et al. (2007) and listed in Chapter 8, to provide the concentration profiles $[C_t]$ for metabolites, viable cell count, and product yield.

TABLE 8.3

Model Predictions of Viable Cell Count and Product Yield Using First-Principle Models Compared to Best-Performing PLS Models Described in Section 8.2.2

			RMSE	
Parameter	Hybrid Model Identifier	Model Description	Train.	Valid.
Viable cell count	A	PLS (on-line) Kontoravdi et al. (2007)	0.33	0.32
	B	PLS (operational param.) Kontoravdi et al. (2007)	0.26	0.17
	C	PLS (on-line) Naderi et al. (2011)	0.49	0.51
	D	PLS (operational param.) Naderi et al. (2011)	0.51	0.28
Product yield	A	PLS (on-line) Kontoravdi et al. (2007)	15.72	19.84
	B	PLS (operational param.) Kontoravdi et al. (2007)	11.56	9.96
	C	PLS (on-line) Naderi et al. (2011)	11.56	10.55
	D	PLS (operational param.) Naderi et al. (2011)	12.56	9.48

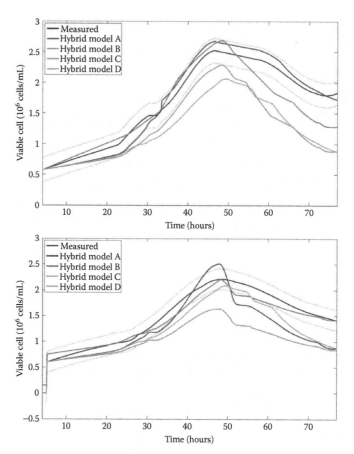

FIGURE 8.8
Prediction of viable cell count for two validation batches using hybrid models described in Section 8.2.3 detailing the structure of the various hybrid models.

8.3 Conclusions

This chapter highlighted the future trends of the (bio)pharmaceutical industry, highlighting the rapidly increasing biologics and monoclonal antibody segment of the market. Regulatory initiatives QbD and PAT (aimed at improving the manufacturing efficiencies through better process and product understanding), structured process development, and more-effective monitoring and control of the processes were also briefly reviewed.

Literature on both first-principles and multivariate approaches to the modeling of pharmaceutical and biopharmaceutical processes indicates a range of challenges for both approaches regarding the development of models capable of capturing the process dynamics and providing an accurate reflection of the process behavior as to the prediction of critical quality attributes and critical process parameters. The benefits of combining the two modeling approaches in a semiparametric hybrid modeling framework were then argued and demonstrated through a case study of monoclonal antibody production using a cell-culture process.

Feature extraction methods such as PARAFAC and PCA were shown to provide a useful tool in exploring process data to gain more in-depth understanding of the correlations, particularly between the operating parameters and process responses. However, to develop models capable of predicting viable cell count, product yield, and glycosylation profiles, hybrid models were shown in this case to be more accurate. Inevitably, the model structures may increase in complexity in such cases, as demonstrated here, but where complex process dynamics, shifts in operating parameters, or other operating strategies make first-principle models and multivariate methods less effective when applied individually, the power of hybrid modeling can lead to significant benefits.

Acknowledgments

The contribution of Dr. Amy Green in terms of the investigations described in the case study detailed in this chapter and carried out within her Engineering Doctorate studies at the Newcastle University, under the supervision of the author, is gratefully acknowledged.

References

Ahn, W. S., J. J. Jeon, Y. R. Jeong, S. J. Lee, and S. K. Yoon. 2008. Effect of culture temperature on erythropoietin production and glycosylation in a perfusion culture of recombinant CHO cells. *Biotechnology and Bioengineering* 101 (6): 1234–1244.

Almquist, J., M. Cvijovic, V. Hatzimanikatis, J. Nielsen, and M. Jirstrand. 2014. Kinetic models in industrial biotechnology: Improving cell factory performance. *Metabolic Engineering* 24: 38–60.

Benyahia, B., R. Lakerveld, and P. I. Barton. 2012. A plant-wide dynamic model of a continuous pharmaceutical process. *Industrial & Engineering Chemistry* 51: 15393–15412.

Brunet, R., G. Guillen-Gosalbez, J. R. Perez-Correa, J. A. Caballero, and L. Jimenez. 2012. Hybrid simulation-optimization based approach for the optimal design of single-product biotechnological processes. *Computers & Chemical Engineering* 37: 125–135.

Carrondo, M., P. Alves, N. Carinhas, J. Glassey, F. Hesse, O. Merten, M. Micheletti, T. Noll, R. Oliveira, U. Reichl, et al. 2012. How can measurement, monitoring, modeling and control advance cell culture in industrial biotechnology? *Biotechnology Journal* 7 (12): 1522–1529.

Dietmair, S., M. P. Hodson, L.-E. Quek, N. E. Timmins, P. Chrysanthopoulos, S. S. Jacob, P. Gray, and L. K. Nielsen. 2012. Metabolite profiling of CHO cells with different growth characteristics. *Biotechnology and Bioengineering* 109: 1404–1414.

Dubey, A., A. U. Vanarase, and F. J. Muzzio. 2012. Impact of process parameters on critical performance attributes of a continuous blender—A DEM-based study. *AIChE Journal* 58: 3676–3684.

Gadgil, M. 2015. Development of a mathematical model for animal cell culture without pH control and its application for evaluation of clone screening outcomes in shake flask culture. *Journal of Chemical Technology and Biotechnology* 90 (1): 166–175.

Geethalakshmi, S., S. Narendran, N. Pappa, and S. Ramalingam. 2012. Development of a hybrid neural network model to predict feeding method in fed-batch cultivation for enhanced recombinant streptokinase productivity in *Escherichia coli*. *Journal of Chemical Technology and Biotechnology* 87: 280–285.

Gerdtzen, Z. 2012. Modeling metabolic networks for mammalian cell systems: General considerations, modeling strategies, and available tools. In *Genomics and Systems Biology of Mammalian Cell Culture*, W. S. Hu and A-P. Zeng (Eds.). pp. 71–108. Berlin, Heidelberg: Springer.

Glassey, J. 2016. Multivariate modelling for bioreactor monitoring and control. In *Bioreactors: Design, Operation and Novel Applications*, C. F. Mandenius (Ed.). London, UK: Wiley-VCH.

Glassey, J., K. Gernaey, C. Clemens, T. Schulz, R. Oliveira, G. Striedner, and C. Mandenius. 2011. Process analytical technology (PAT) for biopharmaceuticals. *Biotechnology Journal* 6 (4): 369–377.

Green, A. 2015. Model based process design for a monoclonal antibody-producing cell line optimisation: Using hybrid modelling and an agent based system. *EngD diss., School of Chemical Engineering and Advanced Materials*, Newcastle University, Tyne, England.

Green, A. and J. Glassey. 2015. Multivariate analysis of the effect of operating conditions on hybridoma cell metabolism and glycosylation of produced antibody. *Journal of Chemical Technology and Biotechnology* 90: 303–313.

Hay, M., D. W. Thomas, J. L. Craighead, C. Economides, and J. Rosenthal. 2014. Clinical development success rates for investigational drugs. *Nature Biotechnology* 32: 40–51.

Herwig, C. 2016. Fixing the negative perception of QbD. *The Medicine Maker,* https://themedicinemaker.com/issues/0815/fixing-the-negative-perception-of-qbd/(accessed May 2017).

Hwang, S., S. Yoon, G. Koh, and G. Lee. 2011. Effects of culture temperature and pH on lag-tagged comp angiopoietin-1 (fca1) production from recombinant CHO cells: Fca1 aggregation. *Applied Microbiology and Biotechnology* 91 (2): 305–315.

Ivarsson, M., T. Villiger, M. Morbidelli, and M. Soos. 2014. Evaluating the impact of cell culture process parameters on monoclonal antibody n-glycosylation. *Journal of Biotechnology* 188: 88–96.

Jang, J. D. and J. P. Barford. 2000. An unstructured kinetic model of macromolecular metabolism in batch and fed batch cultures of hybridoma cells producing monoclonal antibody. *Biochemical Engineering Journal* 4 (2): 153–168.

Kelley, B. 2009. Industrialization of mAb production technology: The bioprocessing industry at a crossroads. *mAbs* 1: 443–452.

Kizhedath, A., S. Wilkinson, and J. Glassey. 2016. Applicability of computational toxicology in monoclonal antibody therapeutics production: Status quo and scope. *Archives of Toxicology.* doi: 10.1007/s00204-016-1876-7.

Konno, Y., Y. Kobayashi, K. Takahashi, E. Takahashi, S. Sakae, M. Wakitani, K. Yamano, T. Suzawa, K. Yano, T. Ohta, et al. 2012. Fucose content of monoclonal antibodies can be controlled by culture medium osmolality for high antibody-dependent cellular cytotoxicity. *Cytotechnology* 64 (3): 249–265.

Kontoravdi, C., D. Wong, C. Lam, Y. Lee, M. Yap, E. Pistikopoulos, and A. Mantalaris. 2007. Modeling amino acid metabolism in mammalian cells toward the development of a model library. *Biotechnology Progress* 23: 1261–1269.

Lakerveld R., B. Benyahia, R. D. Braatz, and P. I. Barton. 2013. Model-based design of a plant-wide control strategy for a continuous pharmaceutical plant. *AIChE Journal* 58: 3671–3685.

Li, J., C. Wong, N. Vijayasankaran, T. Hudson, and A. Amanullah. 2012. Feeding lactate for CHO cell culture processes: Impact on culture metabolism and performance. *Biotechnology and Bioengineering* 109 (5): 1173–1186.

Mercier, S., B. Diepenbroek, M. Dalm, R. Wijffels, and M. Streefland. 2013. Multivariate data analysis as a PAT tool for early bioprocess development data. *Journal of Biotechnology* 167: 262–270.

Naderi, S., M. Meshram, C. Wei, B. McConkey, B. Ingalls, H. Budman, and J. Scharer. 2011. Development of a mathematical model for evaluating the dynamics of normal and apoptotic Chinese hamster ovary cells. *Biotechnology Progress* 27 (5): 1197–1205.

Nolan, R. P. and K. Lee. 2011. Dynamic model of CHO cell metabolism. *Metabolic Engineering* 13: 108–124.

Ollmann Saphire, E., P. W. H. I. Parren, R. Pantophlet, M. B. Zwick, G. M. Morris, P. M. Rudd, R. A. Dwek, R. L. Stanfield, D. R. Burton, and I. A. Wilson. 2001. Crystal structure of a neutralizing human IGG against HIV-1: A template for vaccine design. *Science* 293 (5532): 1155–1159.

Ozturk, S., M. Riley, and B. Palsson. 1992. Effects of ammonia and lactate on hybridoma growth, metabolism, and antibody production. *Biotechnology and Bioengineering* 39 (4): 418–431.

Pais, D. A. M., M. J. Carrondo, P. M. Alves, and A. P. Teixeira. 2014. Towards real-time monitoring of therapeutic protein quality in mammalian cell processes. *Current Opinion in Biotechnology* 30C: 161–167.

Pharma. 2016. World preview 2016, outlook to 2022, evaluate pharma. http://info.evaluategroup.com/rs/607-YGS-364/images/wp16.pdf (accessed May 2017).

Popp, O., D. Müller, K. Didzus, W. Paul, F. Lipsmeier, F. Kirchner, J. Niklas, K. Mauch, and N. Beaucamp. 2016. A hybrid approach identifies metabolic signatures of high-producers for Chinese hamster ovary clone selection and process optimization. *Biotechnology and Bioengineering* 113 (9): 2005–2019.

Rantanen, J. and J. Khinast. 2015. The future of pharmaceutical manufacturing sciences. *Journal of Pharmaceutical Sciences* 104: 3612–3638.

Rathore, A. S. and H. Winkle. 2009. Quality by design for biopharmaceuticals. *Nature Biotechnology* 27 (1): 26–34.

Rogers, A., A. Hashemi, and M. G. Ierapetritou. 2013. Modeling of particulate processes for the continuous manufacture of solid-based pharmaceutical dosage forms. *Processes* 1: 67–127.

Selvarasu, S., Y. S. Ho, W. P. K. Chong, N. S. C. Wong, F. N. K. Yusufi, Y. Y. Lee, M. G. Yap, and D. Y. Lee. 2012. Combined in silico modeling and metabolomics analysis to characterize fed-batch CHO cell culture. *Biotechnology and Bioengineering* 109: 1415–1429.

Sen, M., D. Barraso, R. Singh, and R. Ramachandran. 2014. A multi-scale hybrid CFD-DEM-PBM description of a fluid-bed granulation process. *Processes* 2: 89–111.

Taschwer, M., M. Hackl, J. A. Hernandez Bort, C. Leitner, N. Kumar, U. Puc, J. Grass, M. Papst, R. Kunert, F. Altmann, and N. Borth. 2012. Growth, productivity and protein glycosylation in a CHO EpoFc producer cell line adapted to glutamine-free growth. *Journal of Biotechnology* 157 (2): 295–303.

Teixeira, A., R. Oliveira, P. Alves, and M. Carrondo. 2009. Advances in on-line monitoring and control of mammalian cell cultures: Supporting the PAT initiative. *Biotechnology Advances* 27 (6): 726–732.

Tharmalingam, T., C.-H. Wu, S. Callahan, and C. T. Goudar. 2015. A framework for real-time glycosylation monitoring (RT-GM) in mammalian cell culture. *Biotechnology and Bioengineering* 112: 1146–1154.

Trummer, E., K. Fauland, S. Seidinger, K. Schriebl, C. Lattenmayer, R. Kunert, K. Vorauer, R. Weik, N. Borth, H. Katinger, and D. Muller. 2006. Process parameter shifting: Part I. Effect of DOT, pH, and temperature on the performance of EpoFc expressing CHO cells cultivated in controlled batch bioreactors. *Biotechnology and Bioengineering* 94 (6): 1033–1044.

U.S. Food and Drug Administration. September 2004. Guidance for industry, process analytical technology—A framework for innovative pharmaceutical development, manufacture and quality assurance. *Pharmaceutical* cGMPS. Rockville, MD: Author.

Velayudhan, A. and M. Menon. 2007. Modeling of purification operations in biotechnology: Enabling process development, optimization, and scale-up. *Biotechnology Progress* 23 (1): 68–73.

von Stosch, M., J.-M. Hamelink, and R. Oliveira. 2016. Hybrid modeling as a QbD/PAT tool in process development: An industrial *E. coli* case study. *Bioprocess and Biosystems Engineering* 39: 773–784.

Wassgren, C. R., B. Freireich, J. F. Li, and J. D. Litster. 2011. Incorporation particle flow information from discrete element simulations in population balance models of mixer-coaters. *Chemical Engineering Science* 66: 3592–3604.

Wu, H., Z. Dong, H. Li, and M. Khan. 2015. An integrated process analytical technology (PAT) approach for pharmaceutical crystallization process understanding to ensure product quality and safety: FDA scientist's perspective, *Organic Process Research & Development* 19: 89–101.

Xing, Z., N. Bishop, K. Leister, and Z. J. Li. 2010. Modeling kinetics of a large scale fed batch CHO cell culture by Markov Chain Monte Carlo method. *Biotechnology Progress* 26 (1): 208–219.

Index

Note: Page numbers followed by f and t refer to figures and tables, respectively.